DOS and Windows Essentials

Gerald J. Lemay
Electrical and Computer Engineering Department
University of Massachusetts Dartmouth

Addison-Wesley Publishing Company

Menlo Park, California · Reading, Massachusetts · New York
Don Mills, Ontario · Harlow, UK · Amsterdam · Bonn · Sydney
Singapore · Tokyo · Madrid · San Juan, Puerto Rico · Paris
Milan · Seoul · Taipei · Mexico City

DOS and Windows Essentials

This is a module in *The Engineer's Toolkit,* an Addison-Wesley SELECT edition. Contact your sales representative for more information.

Photo Credits:
Chapter 1: Courtesy of Unisys
Chapter 2: © Laima Druskis Photography/Jeroboam, Inc.
Chapter 3: Courtesy of International Business Machines Corporation
Chapter 4: Courtesy of Microsoft Corporation

ISBN 0-8053-6372-6

Addison-Wesley Publishing Company
2725 Sand Hill Road
Menlo Park, CA 94025

Contents

1

An Introduction to User Environments

First Generation Computers

In 1951 the UNIVAC calculated the U.S. census and established the first generation of computers. These early machines were big, expensive, and required a climate-controlled environment to compensate for their numerous heat-generating vacuum tubes. They also required a human operator to manually load and run programs and data. Operating systems, which perform these tasks automatically, had not yet been developed. This chapter examines operating systems and operating environments, both of which are designed to help computer users run programs and manage files.

INTRODUCTION

This chapter introduces you to the MS-DOS (DOS) operating system and the Windows operating environment. You first examine DOS's traditional command-line interface, as well as its newer desktop interface. Then you explore Windows' user-friendly graphical interface.

1-1 USING OPERATING SYSTEMS

In this section, you will learn how operating systems manage a computer's resources.

A *computer* is a machine designed to perform operations specified by a set of instructions called a *program*. Computer *hardware* refers to the computer equipment, and computer *software* refers to the programs that describe the steps you want the computer to perform.

There are four primary hardware components of a computer system:

- The *central processing unit* (CPU) executes program instructions by performing arithmetic and logical operations on data. Programs and program data needed by the CPU must be organized into computer-readable documents called *files* and be temporarily stored in *memory*.
- *Input devices* accept data in a form the computer can use and send the data to the central processing unit. Examples of input devices are the keyboard, mouse, and trackball.
- *Output devices* display processed data. Examples of output devices are the monitor (or screen) and printer.
- *Secondary storage devices* store program and data files permanently. Examples of secondary storage devices are diskettes, tapes, and CD-ROM disks.

Figure 1-1 shows a typical computer system.

Figure 1-1
A Computer System

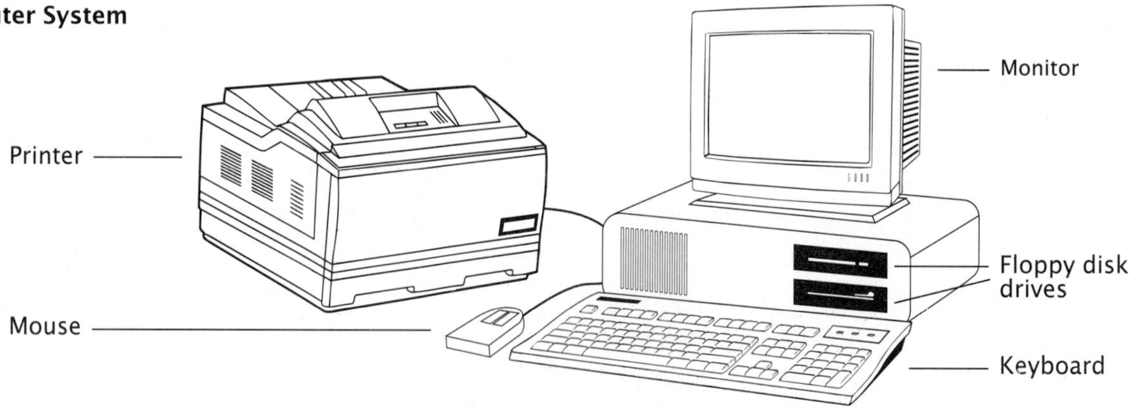

Two categories of computer software are applications programs and operating systems. *Applications software,* such as word processors and spreadsheets, perform specific tasks and functions. An *operating system* is a set of programs that enables a computer to manage its hardware and software. The operating system used with IBM personal computers and IBM compatibles is *MS-DOS* (or *DOS*), which stands for Microsoft Disk Operating System. IBM's version of DOS, called PC-DOS, was first made available for personal computers in 1981.

When you turn on the computer, DOS is loaded into the computer's memory. This process is called *booting.* The DOS *command line* displays the DOS *system prompt,* which typically appears as C> or A>. DOS uses a *command-line interface* in which you tell the computer what you want done by typing commands at the system prompt.

In Chapter 2, you will learn how to start DOS, use basic DOS commands to manage files, and use online Help to obtain more information about DOS.

1-2 USING OPERATING ENVIRONMENTS

One limitation of DOS's command-line interface is that it doesn't help you determine which of the more than 100 DOS commands you should use and how the commands should be entered. In contrast *operating environments* simplify the interface between the user and the computer's operating system.

DOS offers a more user-friendly interface called the *DOS Shell,* which replaces the command prompt with lists of commands called *menus.* Instead of typing the command name, you just select the command using the keyboard or a mouse. The DOS Shell is shown in Figure 1-2.

The DOS Shell is an example of a graphical user interface. In a *graphical user interface* (GUI), the screen background is displayed as a *desktop.* Programs and documents are displayed in *windows,* rectangular areas on the desktop. Instead of typing hard-to-remember commands, you can select commands from menus or use a mouse to point to *icons,* graphical images that represent items such as programs and documents.

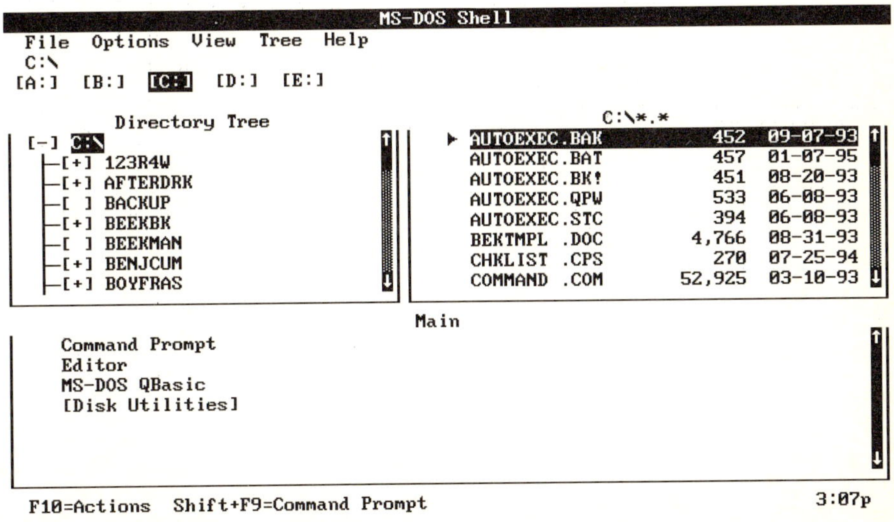

Figure 1-2
The DOS Shell

Figure 1-3
The Windows Desktop

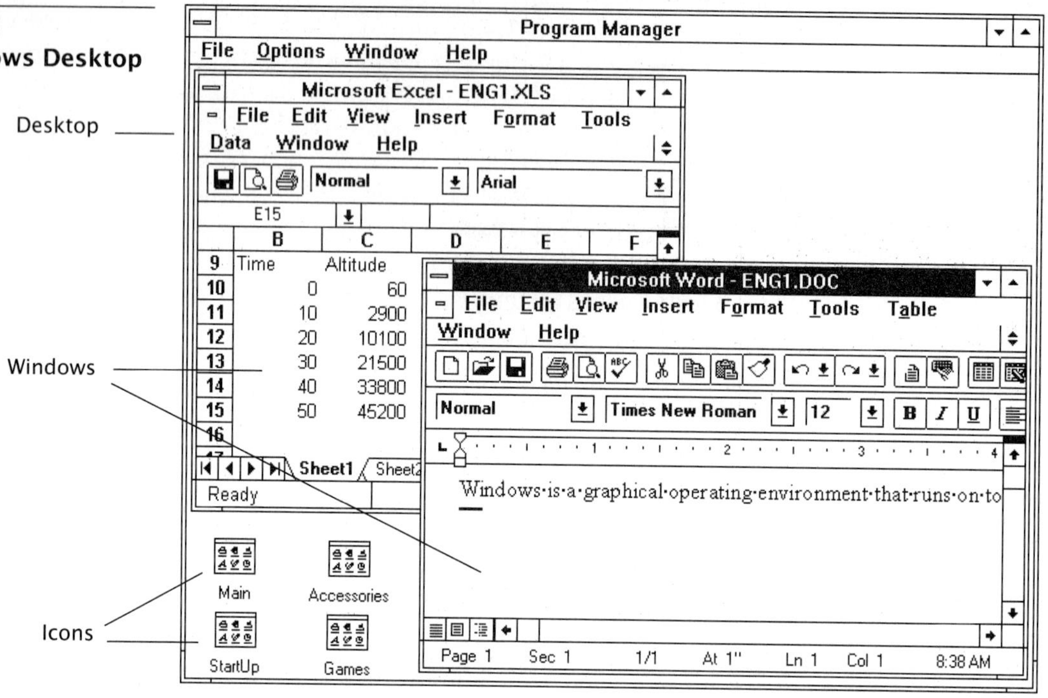

In 1985, Microsoft released a separate graphical user interface for DOS called *Windows*. The Windows desktop is shown in Figure 1-3.

Windows is not an operating system. Instead, it runs in conjunction with DOS and provides features not available with DOS. In addition to its graphical user interface, Windows provides *multitasking* capabilities that enable you to run more than one program simultaneously. Many applications software programs are written to be used specifically with Windows. Windows applications not only have similar interfaces but also enable you to share data between them. In addition, Windows includes its own useful accessories, such as Write, Paintbrush, Calculator, Calendar, and Cardfile. In Chapter 3, you will learn how to start Windows, use basic Windows features, and obtain online Help about Windows. Chapter 4 describes how to use some Windows accessories.

SUMMARY

This chapter introduced you to the DOS operating system and Windows operating environment. You learned how DOS can be used to manage a computer's resources and how Windows extends the capabilities of DOS. In the next two chapters, you will learn how to use DOS and Windows to run programs and manage files.

Key Words

applications software
booting
central processing unit (CPU)
command line
command-line interface
computer
desktop
DOS

DOS Shell
files
graphical user interface (GUI)
hardware
icons
input devices
memory
menus
multitasking
operating environments
operating system
output devices
program
secondary storage devices
software
system prompt
window
Windows

2 Fundamentals of DOS

Command-driven Operating Systems

Whenever your computer is turned on, the operating system is working to manage the computer's resources, working as an interface between you and the details of the machine's operation. Since operating systems are so closely aligned with the computer's hardware, the first operating systems were machine specific. Eventually versions of popular operating systems became available which could be used on different types of computers. This chapter describes how to use the basic commands of one of these operating systems: MS-DOS.

INTRODUCTION

In this chapter, you will learn how to start DOS and use DOS commands to manage files. You will be introduced to DOS conventions for naming files and organizing files into directories. You will also learn how to access online help to obtain explanations and examples of DOS commands.

2-1 GETTING STARTED

While some computers start DOS automatically when they are turned on, you may find it necessary to start DOS manually. In this section, you will learn how to start DOS and format a floppy diskette.

Starting DOS

After you turn on your computer, DOS displays several messages indicating that it is initializing the system; then you should see the system prompt C:\> on the screen. However, if your system displays a screen with the words *MS-DOS Shell* at the top instead of displaying the system prompt, you need to exit this menu-based screen by pressing F3. (Note that some of the keys on the keyboard are labeled F1, F2, and so on. These are *function keys* and have special purposes for some programs.) Likewise, if your computer is connected to a local area network (LAN), you will first need to select the option entitled *Exit to DOS* before you see the system prompt.

Storing Data

Your computer system has two types of storage: primary and secondary. *Primary storage,* also called computer memory, includes RAM and ROM. *RAM* (random access memory) holds instructions and data when the computer is running. RAM storage is temporary. When the computer is off, this memory is usually cleared. *ROM* (read-only memory) can be read from but not written to. Unlike RAM, this memory remains intact whether the computer is on or off. ROM is useful for storing the basic startup instructions that the computer needs when it is first turned on.

Secondary storage is used to store data permanently. A typical computer has several disk drives used for secondary storage. The *hard disk,* typically referred to as drive C:, is permanently housed in its drive. In contrast, *floppy diskettes* can be removed from their disk drives, typically referred to as A: or B:. Currently, two sizes of floppy diskettes are widely used: 3-1/2" and 5-1/4". If you have more than one hard disk or your computer is attached to a network, you may have additional drive letters such as D: and E:.

Data is stored in units of measure called *bytes,* where a byte stores a single character. A *K* (kilobyte) is 1024 bytes, and a *MB* (megabyte) is 1024K. Related data stored on a diskette is organized into *files,* which are then organized into *directories.* Directories can be organized into a directory structure, in which directories are divided into *subdirectories.* The specific directory you are working in at any time is referred to as the *current directory.*

When working with DOS, it is helpful to set the system prompt so that it displays the current directory. To display the current directory, you can type the command

```
PROMPT $P$G
```

and press ENTER.

If the screen shows C:\> your location is the *root directory*. The root directory is the initial or top-level directory from which all the other directories branch out and is represented by a solitary backslash (\). To change to the root directory, you type

```
CD\
```

and press ENTER.

The CD (Change Directory) command followed by the backslash tells DOS to make the root directory the current directory.

Formatting Diskettes

Diskettes must be *formatted* before you can use them. Some diskettes are preformatted for DOS. When you use the format command, any data previously stored on that disk is lost. To format a diskette in drive A:, you type

```
FORMAT A:
```

and press ENTER.

DOS then prompts you to insert a diskette into the floppy disk drive, as shown in Figure 2-1. After you insert the diskette into the drive, you need to press ENTER once more to start the formatting process. DOS may then prompt you to enter an optional *label* of up to 11 characters to name your diskette. If you don't want to provide a label, just press ENTER. DOS then prompts you if you want to format another diskette. To end the formatting process, type NO and press ENTER.

Many of the exercises in this module require the use of a working diskette. In the following exercise, you will create the working diskette from files supplied by your instructor. Although the exercise assumes your instructor will supply files to you on a floppy diskette, in some lab settings the files might be copied from the hard drive or a network.

See your instructor for guidance in such cases. Note that you will be using the DISKCOPY command to create the working diskette.

Figure 2-1
Prompt to Insert Diskette

```
C:\>format a:
Insert new diskette for drive A:
and press ENTER when ready...
```

Try It Prepare a diskette for subsequent exercises in this module.

- Confirm that you are in drive C by typing

```
C:
```

and then press (ENTER).

- Make the root directory the current directory.

The instructions that follow assume that your floppy drive is designated as drive A:. If your floppy drive is drive B:, substitute that letter for the A: used in the following discussion.

- Format a new diskette in drive A:.
- Insert the diskette to be formatted into the floppy drive, and follow the directions displayed on the screen.
- When asked for the label name, type

```
MYDISK
```

and press (ENTER).

- If the files are supplied on the hard drive or a network, see your instructor for specific instructions on how to copy the files to your working diskette. Otherwise, copy the files from the diskette supplied by your instructor by typing

```
DISKCOPY
```

and then press (ENTER).

The source diskette is the diskette supplied by your instructor, and the destination diskette is your MYDISK diskette.

2-2 USING FILE COMMANDS

In this section, you will learn how to copy, view, rename, compare, delete, and undelete files. When you create a file, you give it a descriptive name called the *filename*. This name can be up to eight characters long and must conform to the following three rules:

1. The filename may contain

   - letters from A through Z (uppercase and lowercase are not distinguished)
   - numbers from 0 through 9
   - and the following special characters:

underscore _	caret ^	dollar sign $
tilde ~	exclamation point !	number sign #
percent sign %	ampersand &	hyphen -
braces { }	at sign @	single quotation mark `
apostrophe '	parentheses ( )	

2. The filename may not contain the following characters:

spaces	comma ,	backslash \
period .		

3. The filename must be unique with respect to the names of other files in the same directory.

Optionally, you can add an *extension* of up to three letters to the filename. Different applications use their own extensions. A period (or dot) separates the filename from the extension. For example, TEMP.TXT has TEMP as the filename and .TXT (for text file) as the extension. The extension often provides a quick indication of the purpose of the file and the software program that created it.

Files are created as the output of applications software. After you create several files, you may need to copy, view, rename, compare, or delete them.

Copying a File

It is a good idea to make a copy of a file for safekeeping. If a file is damaged or lost, or if changes made to a file do not satisfy your expectations, you can easily return to the original file. Copying a file makes a duplicate of the original file but does not change it. The general form of the COPY command is

COPY *source destination*

where the *source* is the original file and the *destination* is the copy. For example, if you type

```
COPY TEMP.TXT TEMP2.TXT
```

and press ENTER, a copy of the original TEMP.TXT file is created and named TEMP2.TXT.

Display a File's Contents

You can use the TYPE command to display the contents of a text file on the screen. The general form of the TYPE command is

TYPE *filename*

where *filename* is the name of the file. To display the contents of the file TEMP2.TXT, you type

```
TYPE TEMP2.TXT
```

and press ENTER.

Attempting to view files other than text files will cause strange characters to appear on the screen.

Renaming a File

Sometimes a more descriptive name for a file is desired after the file has already been created. You can use the REN (Rename) command to change the name of a file. The general form of the REN command is

REN *oldname newname*

where the name of the file is changed from *oldname* to *newname*. For example, if you type

```
REN TEMP1.TXT TEMP2.TXT
```

and press (ENTER), the name of the file TEMP1.TXT is changed to TEMP2.TXT.

Comparing Files

Sometimes you want to verify that the contents of two files are identical, such as when you copy files. The general form of the FC (File Compare) command is

FC *filename1 filename2*

which compares the contents of *filename1* with *filename2*.

Deleting a File

When you no longer need a file, you can use the DEL (Delete) command to remove the file from a diskette. The general form of the DEL command is

DEL *filename*

To avoid accidentally deleting the wrong file, you can use the verification option. If you type

```
DEL FILENAME /P
```

and press (ENTER), you will be prompted to verify that you want the file to be deleted.

Undeleting a File

If you accidentally delete a file, you can use the UNDEL (Undelete) command to recover the file. The general form of the UNDEL command is

UNDEL *filename*

To make sure that UNDEL successfully recovers your file, you should use it immediately after having deleted the file.

Try It Try the file commands with your MYDISK diskette.

- Insert your MYDISK diskette into drive A.
- Change to drive A.

- Change to the COM subdirectory on drive A.
- Copy the FORMAT.COM file to a file named FORM.COM.
- Rename the FORMAT.COM file to FOR.COM.
- Compare the FORMAT.COM and FOR.COM files.
- Delete the FOR.COM file.
- Remove the diskette from drive A.

2-3 WORKING WITH DIRECTORIES

Directory commands help you manage your files more efficiently. In this section, you will learn how to change the current directory as well as create, view, and remove directories.

Changing the Current Directory

If you are planning a project that uses two types of files, such as data files and report files, you might create a separate directory for the data files and another directory for the report files. You can use the CD (Change Directory) command to switch between these two directories. You can also use the CD command to display the name of the current directory.

The general form for the CD command is

CD [*drive*:][*path*]

where the text inside the brackets [] is optional. Typing `CD` without parameters displays the current drive and directory.

The *path* is defined as the location of a file with respect to a directory. In Figure 2-2, COM is a subdirectory in the TEMP subdirectory on drive A, and TREE.COM is a file in the COM subdirectory.

Figure 2-2
Sample Directory Structure

```
A:.
└──TEMP
    ├──COM
    │      EDIT.COM
    │      FORMAT.COM
    │      HELP.COM
    │      TREE.COM
    │
    └──HLP
           EDIT.HLP
           DOSHELP.HLP
           MSAV.HLP
           MEMMAKER.HLP
           SMARTMON.HLP
```

When working with files, you can use the full path or a relative path. The *full path* is the location of a file with respect to the root directory. For the TREE.COM file, the full path is A:\TEMP\COM\TREE.COM. The backslashes are used to separate subdirectory names and filenames.

The *relative path* is the location of a file with respect to the current directory. For example, if the current directory is A:\TEMP\COM, then the relative path for the TREE.COM file is simply TREE.COM. DOS lets you specify a path length of up to 63 characters, including backslashes.

The *parent* directory is the previous level in a directory structure. You can change to the parent directory if you type

```
CD ..
```

and press ENTER.

Try It Try the CD command with the directories on the MYDISK diskette.

- Change to drive A.
- Use the relative path to change to the TEMP directory on drive A.
- Use the full path to change to the HLP directory on drive A.
- Return to the parent directory TEMP.

Creating a Directory

In this section, you will learn how to create directories and subdirectories for your files using the MD (Make Directory) command. The general form of the MD command is

MD [*drive*:]*path*

The directories in the WINDOWS directory in the C drive are shown in Figure 2-3. Note that MSAPPS is a subdirectory of the WINDOWS directory, and MSAPPS is the parent directory of nine subdirectories.

Figure 2-3
Windows Directory Structure

```
C:\WINDOWS>tree
Directory PATH listing
Volume Serial Number is 1935-1B04
C:.
├───SYSTEM
├───TEMP
├───MSAPPS
│   ├───WORDART
│   ├───PROOF
│   ├───GRPHFLT
│   ├───TEXTCONV
│   ├───MSGRAPH5
│   ├───ARTGALRY
│   ├───MSINFO
│   ├───SHEETCNV
│   └───ORGCHART
├───TRAVSOFT
└───ORIG
```

Try It Create a subdirectory in the TEMP subdirectory.

- Change to drive A.
- Change to the TEMP subdirectory on drive A.
- Create a subdirectory called TXT.
- Return to the root directory on drive A.
- Use the full path to create an additional subdirectory called RPT in the TEMP subdirectory.

Viewing a Directory

You can use the DIR and TREE commands to view the files and subdirectories on a diskette. The DIR (Directory) command displays a directory listing of the files and subdirectories contained in the current directory. The general form of the DIR command is

DIR [*drive*:][*path*][*filename*]

Several switches are available with the DIR command. For example, the /W (Wide) switch displays a wide version of the directory list, the /P (Pause) switch displays one screen of data at a time, and the /O (Order) switch displays the directory list in alphabetic order. Figure 2-4 shows the directory listing of the TEMP directory.

Figure 2-4
Directory Listing of the Temp Directory

```
A:\TEMP>dir

 Volume in drive A has no label
 Directory of A:\TEMP

.            <DIR>        06-16-94   3:24p
..           <DIR>        06-16-94   3:24p
COM          <DIR>        06-16-94   3:24p
HLP          <DIR>        06-16-94   3:24p
        4 file(s)              0 bytes
                      1,451,520 bytes free
```

You can display the files in the HLP subdirectory by making HLP the current directory and typing `DIR`. If you use the /W switch and type `DIR/W`, DOS displays the directory listing horizontally, as shown in Figure 2-5.

Figure 2-5
Displaying the Files Horizontally

```
A:\TEMP\HLP>dir/w

 Volume in drive A has no label
 Directory of A:\TEMP\HLP

[.]             [..]            EDIT.HLP        DOSHELP.HLP     MSAV.HLP
MEMMAKER.HLP    SMARTMON.HLP
        7 file(s)            135 bytes
                      1,451,520 bytes free
```

For long directory lists, you could use the /P switch to display one screen of the directory list at a time. You are then prompted to `Press any key to continue`. Options for the DIR command may be combined. For example, DIR/O/P lists the directory alphabetically and pauses after each screen.

You can specify groups of files by using *wildcards*, which are special characters that represent one or more characters. The asterisk (*) represents one or more characters and the question mark (?) represents a single character. Here are several examples.

Example	**Displays**
`DIR*.COM`	All files with the COM extension
`DIR.EDIT*`	All files named EDIT with any extension
`DIR MO?E.COM`	All files with the COM extension that have four-character filenames that begin with MO and end with E

The other command for viewing directories is the TREE command. The general form of the TREE command is

TREE [*directory*]

The TREE command includes a /F (Full) switch that you can use to display the names of the files included in each directory. Figure 2-6 shows the results of using the command TREE \F A:

Figure 2-6
Using Tree to View a Directory

```
C:\>tree /f a:
Directory PATH listing
A:.
└───TEMP
    ├───COM
    │       EDIT.COM
    │       FORMAT.COM
    │       HELP.COM
    │       TREE.COM
    │
    └───HLP
            EDIT.HLP
            DOSHELP.HLP
            MSAV.HLP
            MEMMAKER.HLP
            SMARTMON.HLP
```

Try It

Try the DIR and TREE commands with your MYDISK diskette.

- Change to drive A.
- Change to the COM subdirectory in drive A.
- Display the files in this directory in horizontal format.
- Determine which files in the HLP subdirectory begin with the letter M.
- Use the TREE command to view the structure of the COM subdirectory.

Removing a Directory

When you no longer require the files stored in a directory, you may want to free some disk space by using the RD (Remove Directory) command. The general form of the command is

RD [*drive:*]*path*

A directory must be empty before you can remove it. To remove all the files in the COM directory, you could type

```
DEL A:\TEMP\COM
```

and press (ENTER).

DOS then gives you a chance to change your mind by prompting you with the question

```
All files in directory will be deleted! Are you sure (Y/N)?
```

Try It

Try the RD command with your MYDISK diskette in drive A.

- Change to drive A.
- Remove the empty TXT subdirectory.

- Since the HLP subdirectory has files in it, delete these files.
- You need to type Y in response to the prompt.
- Now, remove the empty HLP subdirectory.
- Verify that the HLP and TXT subdirectories have been removed using the TREE command.

2-4 GETTING ONLINE HELP

DOS is a command-based program in which the commands have unique wording and grammar. It is convenient to have a readily accessible reference to the correct form of the commands. DOS provides two kinds of online help: MS-DOS Help and FastHelp.

Using MS-DOS Help

To start MS-DOS Help, type

```
HELP
```

and press ENTER.

The MS-DOS Help Command Reference shown in Figure 2-7 appears.

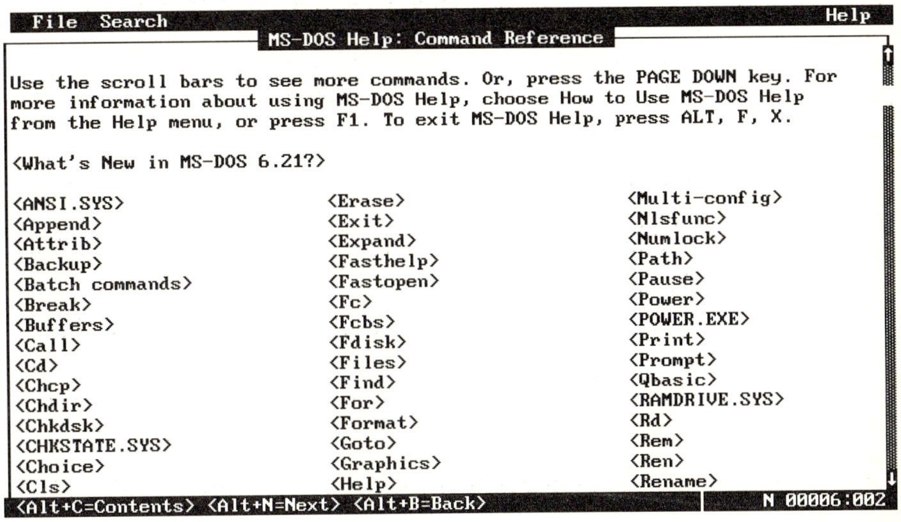

Figure 2-7
MS-DOS Help Topics

Choosing a Topic You can use ↑ and ↓ to vertically move the cursor (highlight) in a column and use TAB to horizontally move the cursor through the list of commands. Pressing the first letter of the command moves the cursor to the first command that begins with that letter, whereas pressing PGDN and PGUP scrolls (quickly moves) through the list of commands. Once the cursor is positioned within the angle brackets < > of the desired command, you can press ENTER to select the command.

Pressing ALT+C displays the MS-DOS Help Contents. Pressing ALT+N displays the next screen, and pressing ALT+B displays the previous screen.

Displaying Information on Commands Most commands have three associated topics. Syntax is provided for all commands and describes how and when to use the command. Notes provides additional information about some commands. Examples also are provided for some of the commands. If you select the DEL command, the screen in Figure 2-8 appears.

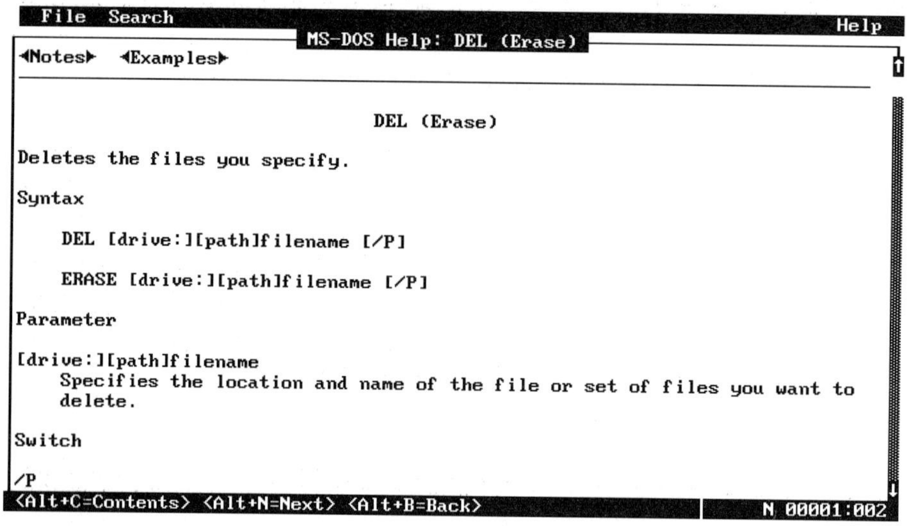

Figure 2-8
Displaying Information About the DEL Command

The Syntax screen provides information about when to use the command and rules for using the command correctly. If the command has associated information, the word Notes is displayed on the top of the screen. To see the notes available for the DEL command, use TAB to position the cursor on Notes and press ENTER. The screen shown in Figure 2-9 appears.

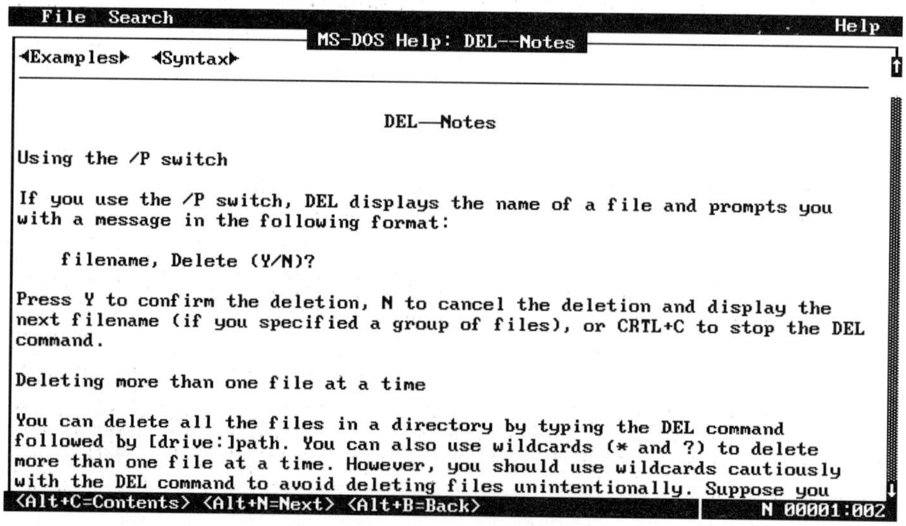

Figure 2-9
DEL Command Notes

If examples showing how to use a command are available, the word Examples is displayed on the top of the screen. To see examples of how to use the DEL command, position the cursor on Examples, and press ENTER. The screen shown in Figure 2-10 appears.

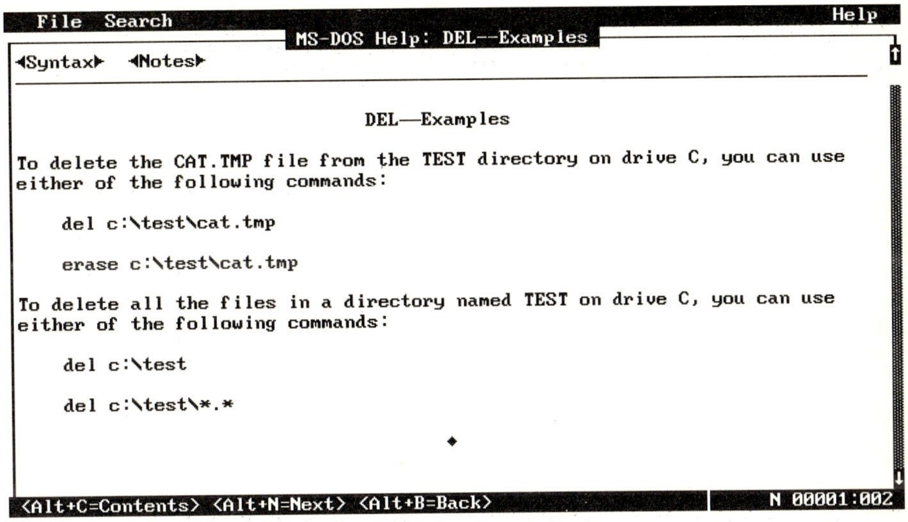

Figure 2-10
DEL Command Examples

Using FastHelp

FastHelp provides an abridged form of online help without leaving the command line. The general form of the FASTHELP command is

FASTHELP [*command*]

If you type

```
FASTHELP DEL
```

and press ENTER DOS displays information about the DEL command. Figure 2-11 shows FastHelp information for the DEL command.

Figure 2-11
The FastHelp Information for the DEL Command

```
C:\>fasthelp del
Deletes one or more files.

DEL [drive:][path]filename [/P]
ERASE [drive:][path]filename [/P]

  [drive:][path]filename  Specifies the file(s) to delete.  Specify multiple
                          files by using wildcards.
  /P                      Prompts for confirmation before deleting each file.
```

If you don't include a command name, a list of DOS comands is displayed with a brief explanation of each.

Try It

Obtain online help about the TREE command.

- Obtain MS-DOS Help about the /A switch of the TREE command.
- Review the examples of the TREE command.
- Exit MS-DOS Help.
- Use FastHelp to get information about the /A switch of the TREE command.

SUMMARY

This chapter introduced you to the MS-DOS operating system and you learned how data is organized into files and directories. You then used DOS commands to create and manage files and directories. You also learned how to access online help to find out more about DOS.

Key Words

bytes
current directory
destination
directories
extension
FastHelp
files
filename
floppy diskettes
formatted
full path
function keys
hard disk
K
label
MB
parent
path
primary storage
RAM
relative path
ROM
root directory
secondary storage
source
subdirectories
wildcards

Exercises

1. Use MS-DOS Help to learn how to change the system prompt to display the current time followed by the greater than (>) symbol.
2. Create a directory on drive A called EX, and copy ANSI.SYS from the DOS directory on drive C to this EX directory.
3. Rename the ANSI.SYS file in the EX directory of drive A to the name ANS.SYS.
4. Remove the directory called EX that you created on drive A in Exercise 2.
5. Display a directory listing of all the files in the DOS directory in wide format with a pause after each screen.
6. Display the directory structure of the root directory of drive C. Do not display all the files.
7. Use MS-DOS Help to obtain information about the CLS command. Write the general form of this command.
8. Use MS-DOS Help to obtain information about the DATE command. Write the general form of this command.
9. Use MS-DOS Help to discover how to display the current time.
10. Display the filenames of all the COM files in the DOS directory of drive C that have M and O as the first two letters.
11. Use the ? wildcard to display the filenames of all the EXE files in the DOS directory of drive C that have five-character filenames starting with S and ending with E.
12. Use the ? and * wildcards to display the filenames of all the files (regardless of extension) in the DOS directory of drive C that have four-character filenames starting with M and ending with V.

3 An Introduction to Windows

Graphical Use Interface

Operating systems changed dramatically with the introduction of graphical user interfaces (GUI) in the early 1980s. No longer were computer users faced with a blank command line they needed to fill with complicated command sequences. Instead, they could use a mouse to select choices displayed for them in a carefully designed graphical environment. The choices might appear as a text menu of available command names, or as icons representing available operations. These graphical operating environments not only made computers easier to use, they also enabled computer users to run several programs simultaneously and to share data between programs. For example, a computer user could load a chart created by a spreadsheet program into a memo created with a word processor and then address the memo to individuals identified by a database program. This chapter describes how to perform some basic computer operations using the Windows operating environment.

INTRODUCTION

As you saw in Chapter 1, Windows is a graphical user interface that runs on top of DOS. In this chapter, you will be introduced to the Windows desktop and learn how to use a mouse to work with Windows. Then you will use Windows to manage files and programs. You also will learn how to access Windows online help.

3-1 WORKING WITH A GRAPHICAL OPERATING ENVIRONMENT

The first thing you notice when you start Windows is that the DOS system prompt is replaced with the Windows graphical user interface. If your computer is using a menu system or the DOS Shell, you can start Windows by selecting the appropriate menu item. To start Windows at the DOS prompt, you type

```
WIN
```

and press (ENTER).

The Windows logo appears, followed by the Windows *desktop*, a graphics-based work area designed to guide you as you use the program. The Windows desktop is shown in Figure 3-1. Applications software and files are located inside rectangular areas called *windows.* You can interact with the computer using lists of commands called menus or graphical images called icons.

In this section, you will start Windows, learn about the different parts of Windows desktop, and interact with Windows using a mouse.

Figure 3-1
The Windows Desktop

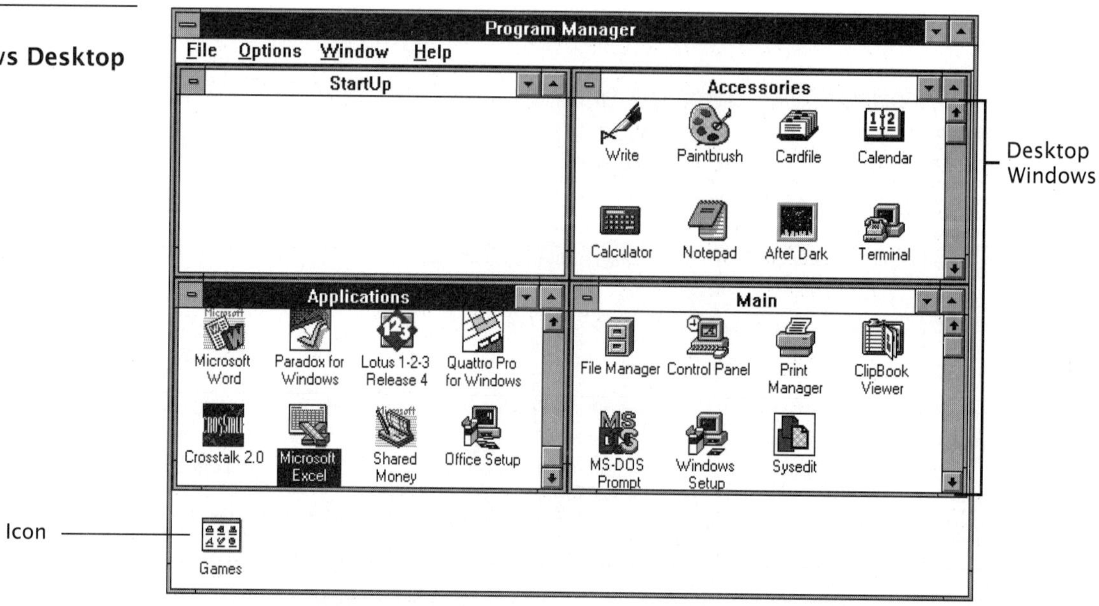

Using a Mouse

You can interact with Windows using the keyboard and the mouse or trackball. The *insertion point* indicates your location in the current document. You can change the location of the insertion point using the mouse. When you move the mouse, the *mouse pointer* follows the movement on the screen. The mouse pointer has different shapes depending on the current activity. Typically it takes the form of an arrow or an I-Beam as shown in Figure 3-2. The basic mouse operations are point, click, double-click, and drag. To *point* to an object, you move the mouse pointer so that it is on the object. To *click*, you press and release the left mouse button. To *double-click*, you click twice in rapid succession. To *drag* an object, you point to the object, hold down the left mouse button, move the mouse pointer to the desired location, and then release the left button. Dragging is useful for moving and sizing an object, and for choosing a command. You *select* (or choose) an object by pointing to the object and clicking the left mouse button.

Figure 3-2
Navigating in a Window

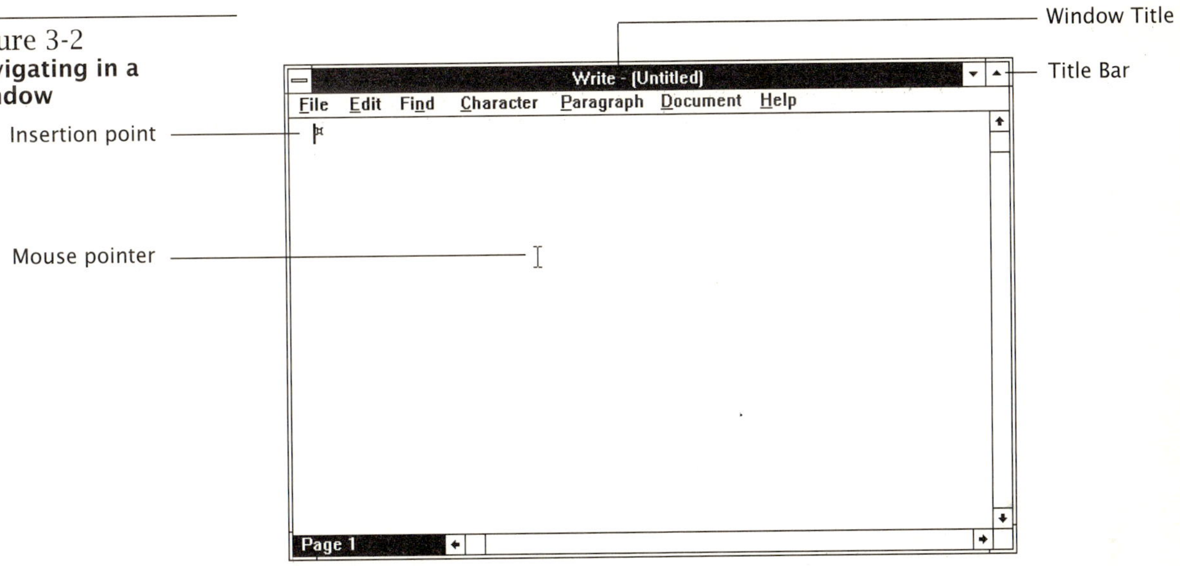

Working with Menus

A *title bar* is located at the top of each window and displays the *window title*. The *menu bar* is located below the title bar and typically includes File, Edit, and Help menus. If you click on a menu item, a *drop-down menu* appears that provides additional choices. You choose the items in a drop-down menu by clicking the option name, as shown in Figure 3-3.

Applications software designed specifically for Windows all use a consistent interface. The order of menu items from left to right is File, Edit, and then the specific menu items for the application. This regular ordering of menu items makes it easier to learn new applications since you usually can apply skills learned in one Windows application to another application.

Manipulating Windows

You can change the size, location, and shape of windows. As shown in Figure 3-3 the top-right corner of each window includes *buttons* you can use

Figure 3-3
A Drop-Down Menu

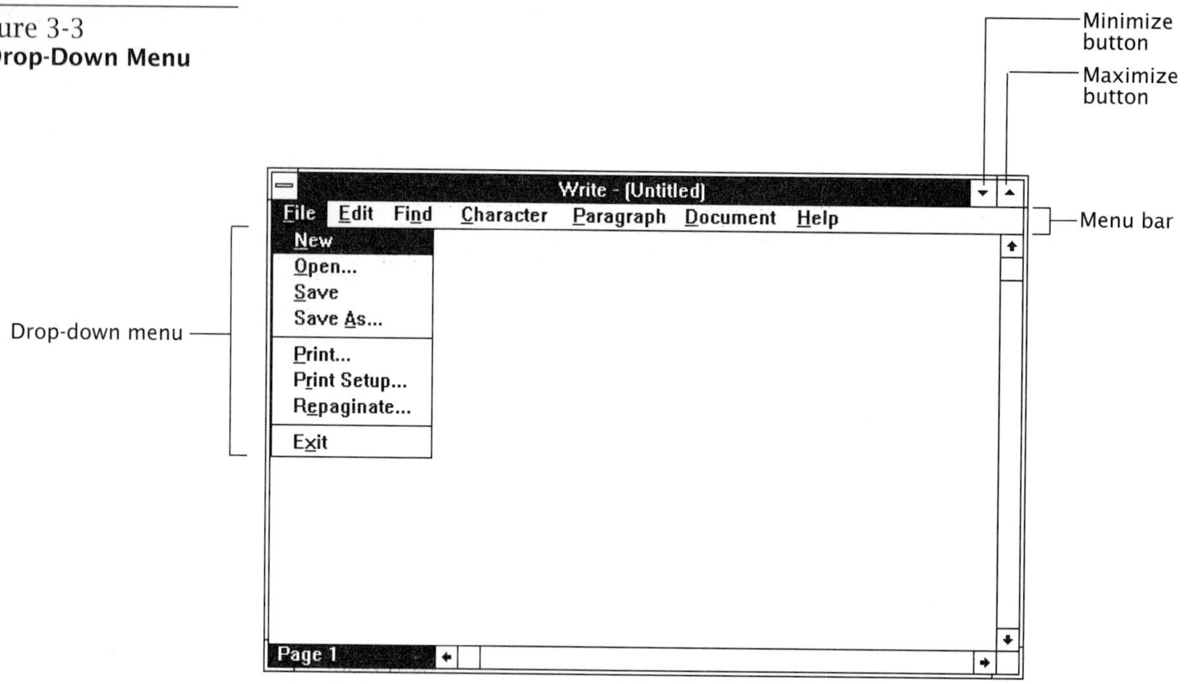

to change the size of the window. You can click the *Maximize button* to expand a window to the full screen. The Maximize button is then replaced with the *Restore button* shown in Figure 3-4, which you can click to return the window to its previous size. If you click the *Minimize button*, the window is reduced to an icon that is placed at the bottom of the screen. You then can return to the application by double-clicking its icon.

Figure 3-4
The Restore Button

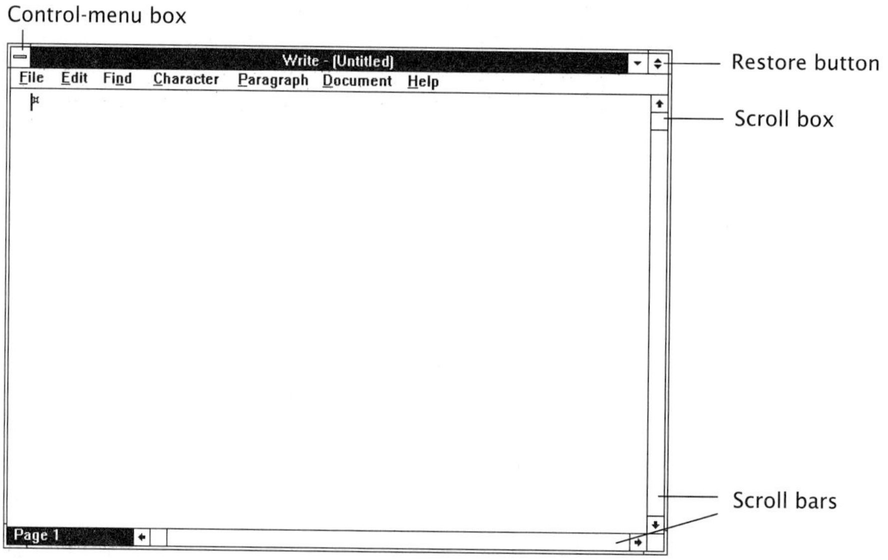

Sometimes the contents of a file exceed the boundaries of the window. In these cases, you can see the contents that are hidden from view by clicking either arrow in the *scroll bar* shown in Figure 3-4 or by dragging the *scroll box* to expand the window.

You also can use the mouse to move a window or change its shape. If you point to a window's title bar and drag the pointer in one direction, the window will move in that direction. If you move the mouse pointer to the edge of a window, the pointer's shape changes. If you then drag the edges, you can change the height or width of the window.

There are two ways you can close a window. From its File menu, you can choose Exit. Or you can double-click the *Control-menu box* shown in Figure 3-4. In the next section, we will discuss the Program Manager window. You close the Program Manager window to exit from Windows.

Try It Use the mouse to manipulate a window.

- Start Windows.
- Increase the size of the Program Manager window.
- Shrink the window to an icon.
- Return to the Program Manager window.
- Drag the Program Manager window to the right.
- Change the shape of the Program Manager window.
- Exit Windows by double-clicking the Control-menu box and responding OK to the prompt `This will end your Windows session.`

3-2 WORKING WITH THE PROGRAM MANAGER

The *Program Manager* shown in Figure 3-5 is started when you run Windows, stays running for the duration of your Windows session, and controls Windows applications. You use the Program Manager to start applications, and organize applications and files. In this section, you will learn about the Program Manager, use the Task List, and access online Help. The appearance of the objects displayed in the Program Manager window depends on the software applications installed on the system. You can change the position of a window or an icon by dragging it to a new location within the Program Manager window.

There are two kinds of windows: applications windows and document windows. An *applications window* contains a running program. Any associated documents are displayed inside *document windows.* When windows are reduced, they change into icons.

Applications can also be displayed as icons in a group window. When group windows are reduced, they change into *group icons.* The Main group window contains the Windows system applications. This group includes File Manager, Control Panel, Print Manager, Clipboard Viewer, DOS prompt, Windows Setup, PIF Editor, and Read Me.

The Accessories group contains Windows programs that support various tasks such as word processing and communications. These accessories include Write, Paintbrush, Cardfile, Calendar, and Calculator. The Applications group contains the applications found on your hard disk during Setup. The Startup group contains the applications that are automatically run when you start Windows. The Games group contains game programs.

Figure 3-5
Group Windows and Group Icon

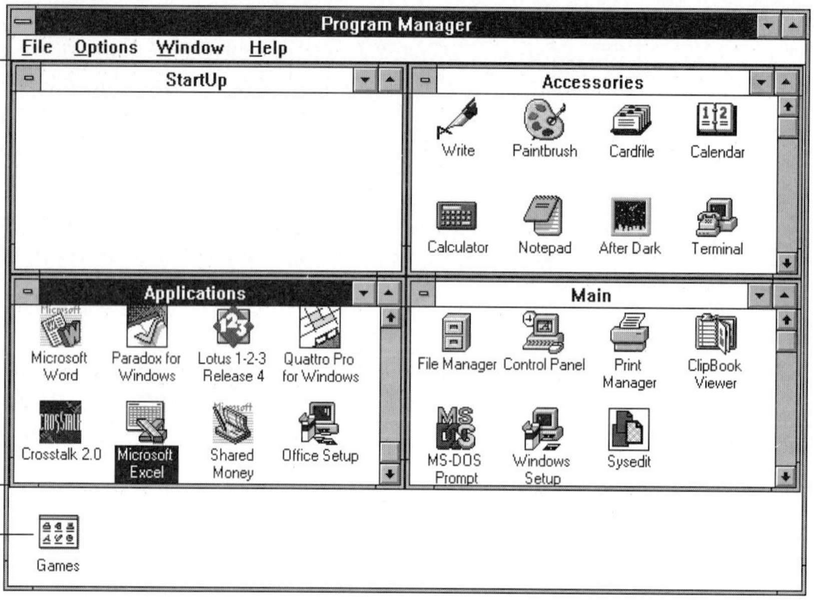

Starting an Application

You can manage your windows and switch from one application to another by using the *Task List.* You can open the Task List, shown in Figure 3-6, by pressing CTRL+ESC or by double-clicking anywhere in the desktop area. Since the Program Manager continues to run during your Windows session, it is always included in the Task List. Opening the Task List is a convenient way to access the Program Manager.

Figure 3-6
The Task List Window

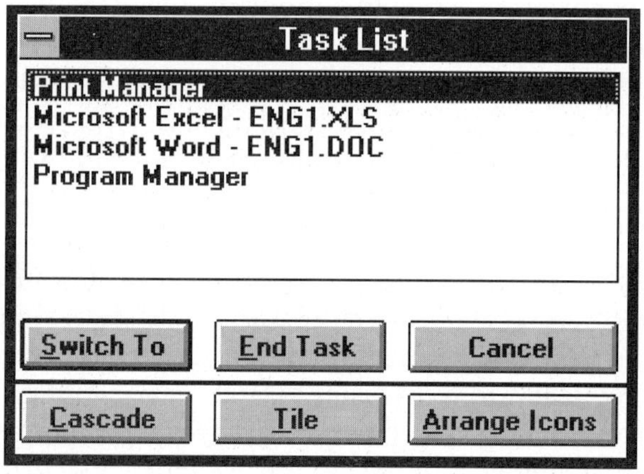

You can switch to another program in the list by double-clicking on that item. If several windows are open at the same time, you can view them more easily by selecting Cascade or Tile. If you select *Cascade*, the windows appear one in front of the other, as shown in Figure 3-7.

You can bring a cascaded window to the front by clicking anywhere in it. If you select *Tile*, the opened windows will appear side by side, as shown in Figure 3-8. You can end a program by choosing that program in the Task List and selecting the End Task option.

Figure 3-7
Cascaded Windows

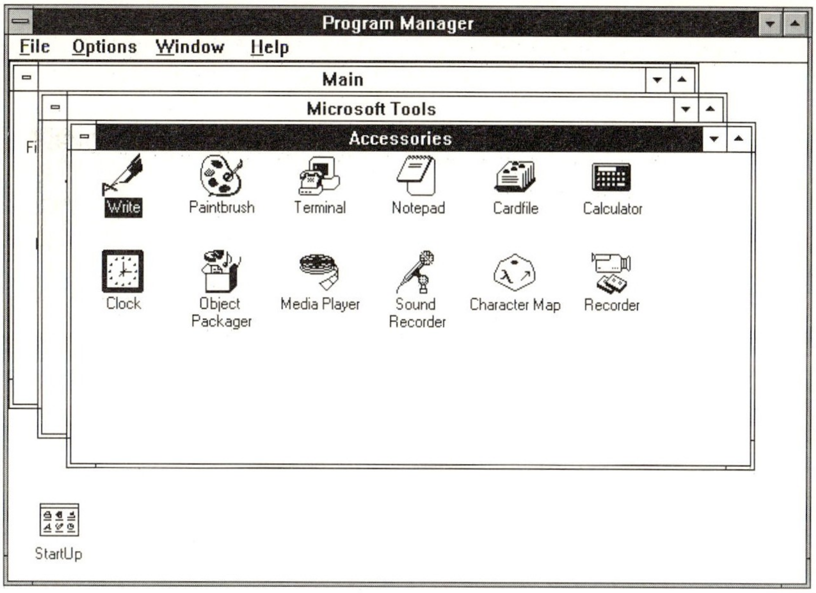

Figure 3-8
Tiled Windows

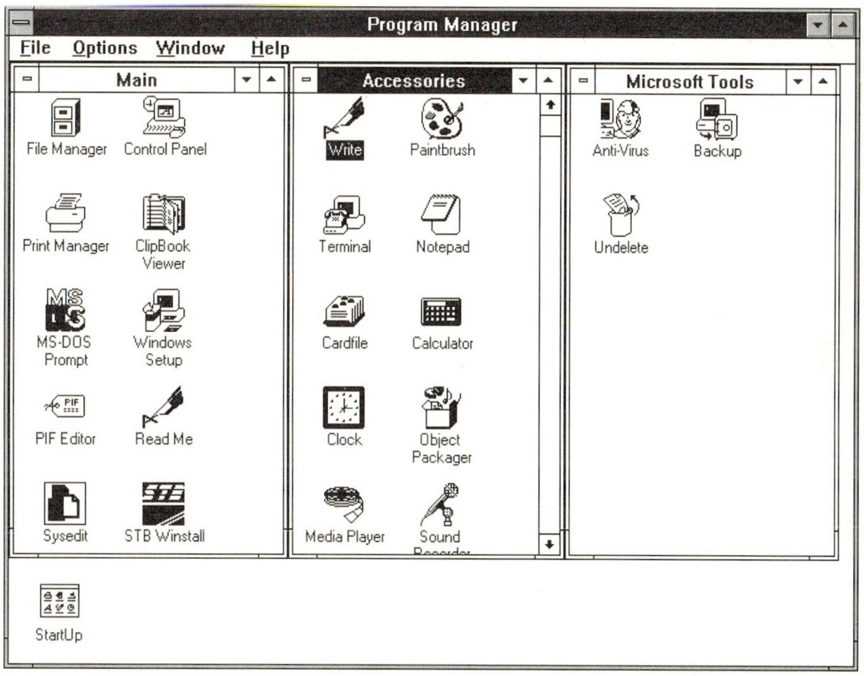

Try It

Use the Task List to complete these exercises.

- Open the Task List.
- Use the Accessories icon to open the Write window.
- Open the Task List and choose Program Manager.
- Select the Accessories icon, and then open the Calendar window.
- Use the Task List to cascade the Write window and the Calendar window. Practice bringing either window to the front by clicking in the desired window.

- Use the Task List to tile the Write window and the Calendar window.
- End the Write task.
- End the Calendar task.

You can also start an application by double-clicking its icon or from the File menu, you can choose Run. Windows then displays a dialog box requesting additional information.

Working with Dialog Boxes

Windows uses *dialog boxes* to prompt you for information as well as supply you with information about a current task. If a menu item is followed by an ellipsis (...), a dialog box appears when you choose that menu item. For example, the Run dialog box shown in Figure 3-9 appears if, from the File menu, you choose Run to start an application.

Figure 3-9
Run Dialog Box and Print Dialog Box

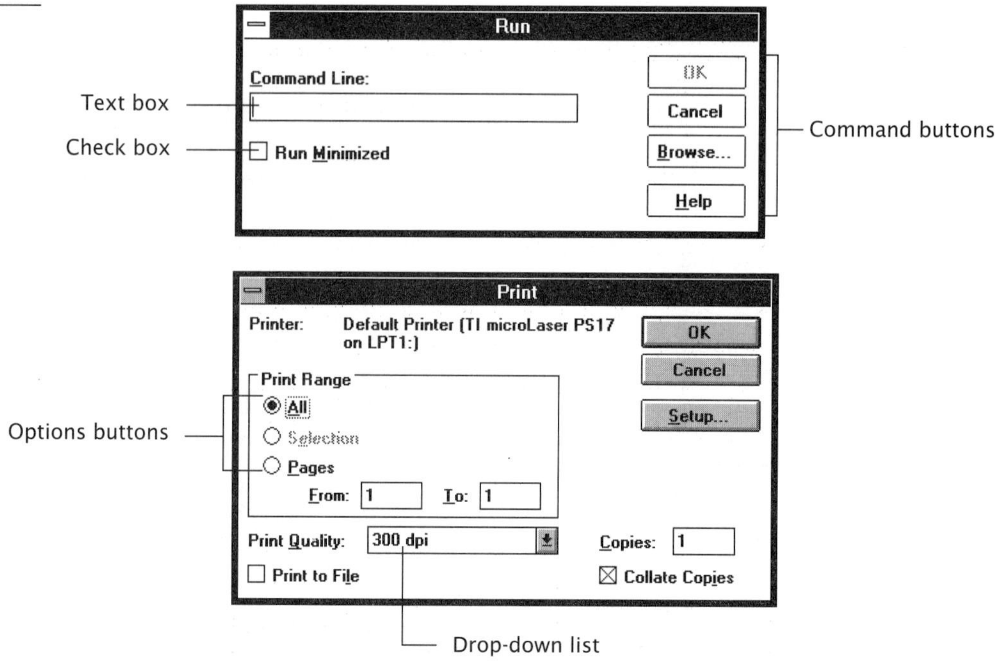

The dialog box contains options you can select to execute the command. You use the *text box* to type in information and *command buttons* to initiate an immediate action. The OK, Cancel, and Help buttons are common to all dialog boxes. You use the *check box* to select or clear the specific option.

The Print dialog box shown in Figure 3-9 contains other dialog box features. You use *options buttons* to select mutually exclusive options. If you check the down arrow in a *drop-down list*, available options are displayed. To find out about these and other dialog box options, you can click the Help button. To close the dialog box, you double-click the Control-menu box.

Using Online Help

Windows provides online help. To obtain information about the Program Manager, from the Program Manager menu, you can choose Help. If you then choose Contents, the topics included in Program Manager Help appear, as shown in Figure 3-10.

Figure 3-10
Program Manager Help Contents

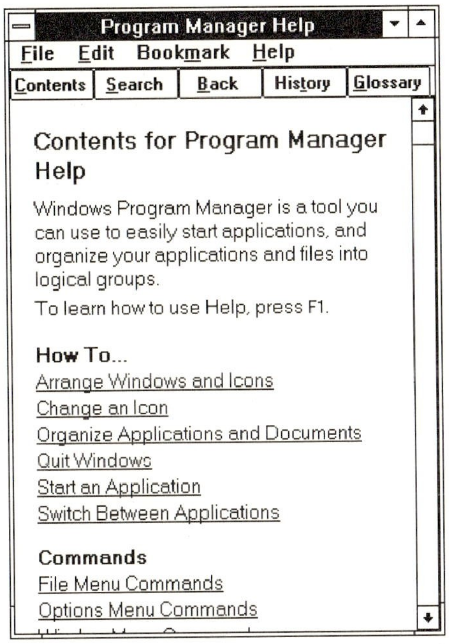

To obtain help on a specific topic, you can choose the Search button. The search screen shown in Figure 3-11 appears. If you type a topic name, Windows displays a list of topics that include the word you typed. If you then highlight the name and click the Show Topics button, a list of subtopics is displayed. To obtain information about one of these subtopics, select the subtopic and click the Go To button. To close the Program Manager Help screen, double-click its Control-menu box.

Figure 3-11
Help Search Window

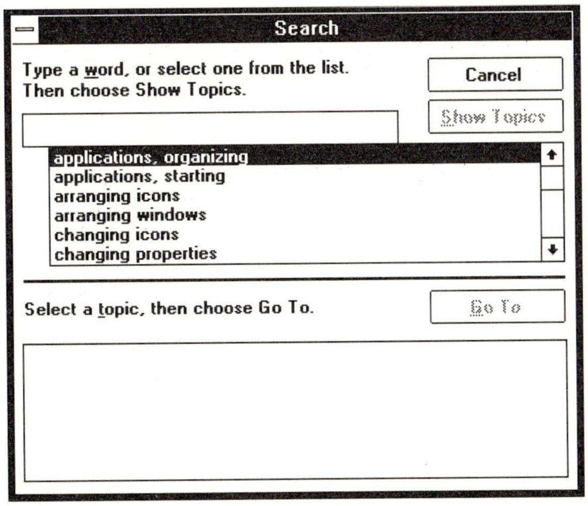

Try It

In this exercise, you will be using the online help to obtain information about the Program Manager and the Run dialog box.

- From the Program Manager menu, choose Help, and then select Contents.
- Search for information about the title bar, icons, and applications.
- Close the Help screen.
- From the File menu, choose Run.
- Use the Help button to find out information about the Run dialog box.
- Close the Help screen and the Run dialog box.

3-3 WORKING WITH THE FILE MANAGER

You use the *File Manager* to organize your files and directories. The File Manager provides a visual representation of the *directory structure* (or organization) in a directory window. A *directory window* displays file and folder icons for directories and files allowing you to perform DOS directory and file commands by clicking and dragging the mouse pointer. In this way, DOS operations are no longer command driven.

To start the File Manager, you double-click on the File Manager icon in the Main group window. A typical File Manager screen is shown in Figure 3-12.

Figure 3-12
File Manager Directory Window

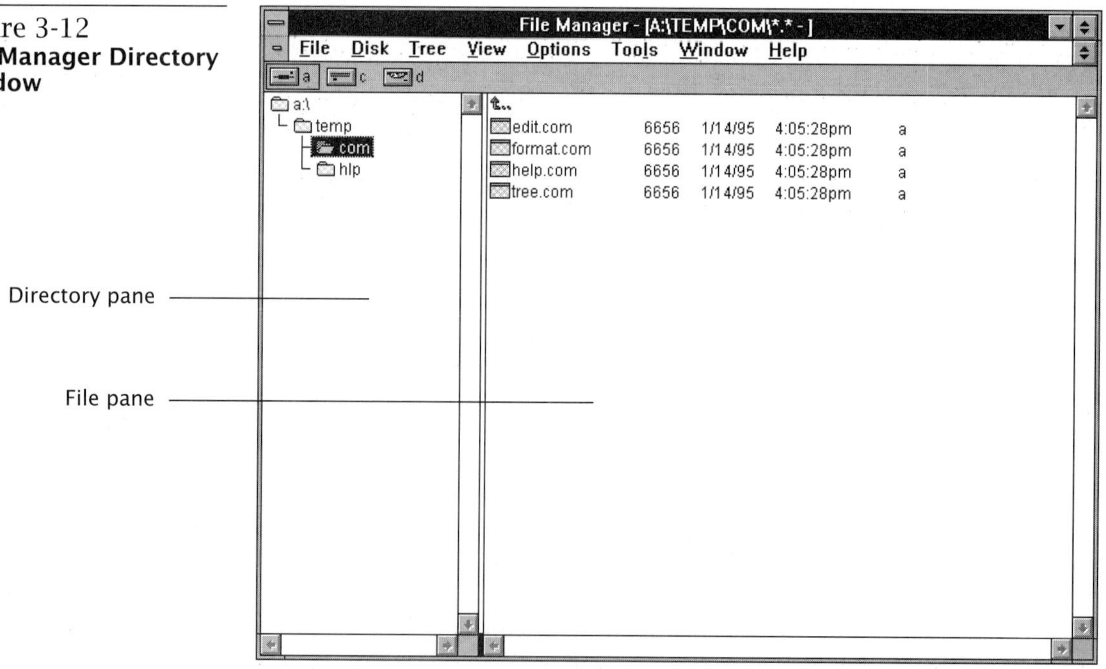

Viewing Files and Directories

You use the View menu to view the directory names. From the View menu, you choose Tree and Directory, as shown in Figure 3-13. The *directory pane*, displayed on the left side of the window, is a directory tree structure similar to that displayed by the DOS TREE command. The *file pane* to the right displays the contents of the selected directory. Folder icons are located next to directory names, and file icons are located next to the filenames. To select a directory, you double-click on the folder icon on the left side of the window. The folder opens, and the contents are displayed on the right side. You can use the scroll bars to see more of the directory structure.

You can display only the directory tree. To do so, from the View menu, you choose Tree Only. To view only the directory contents, from the View menu, choose Directory Only. To move the screen partition so that you can view more of one side of the screen, from the View menu, choose Split. You then can drag the vertical line to the desired position and click.

Figure 3-13
The View Menu

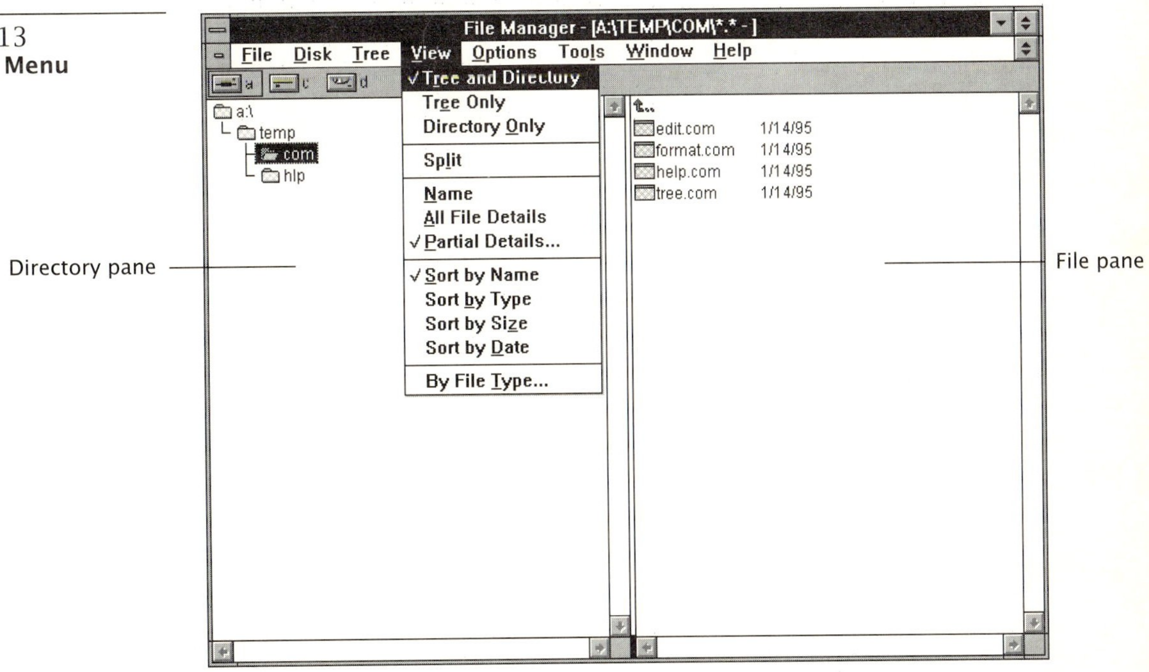

You can view all or some of the file information such as the size, the date of the last modification, the time of the last modification, and the attributes. To see partial details, from the View menu, choose Partial Details.... Then use the Partial Details dialog box to specify the display. To see only the filenames, from the View menu, choose Name. To see everything, from the View menu, choose All File Details.

To sort the files by name, type, size, or date, from the View menu, choose the appropriate Sort command. You then use the Sort dialog box to specify how you want the files sorted.

You can use wildcards to display all files with similar filenames or extensions. From the View menu, choose By File Type.... Then use the By File Type dialog box to specify the display. For example, to display all the filenames that have the extension .COM, you can type `*.COM`. To display all filenames, you can type `*.*`.

Manipulating Files and Directories

You can use the mouse to copy one or more files. First make sure the destination is visible, where the destination is either a directory or drive icon. Then select the source files you want copied in the file pane. To select a continuous range of files, hold down SHFT and click the desired file icons. To select noncontiguous files, hold down CTRL while you click the desired file icons. When the files are selected, hold down CTRL while you use the mouse to drag the source file icons to the destination. To complete the operation, you will have to respond to a confirmation prompt.

You also can move one or more files. Again, make sure the destination icons are visible. Then select the files as in the copy command but instead of pressing CTRL and dragging, press SHFT and drag the file(s) to their new location. You'll also have to respond to the confirmation prompt.

To delete one or more files, select the file icons as described. Then press DEL and respond to the confirmation prompt.

You need to use the File menu to rename a file. From the File menu, choose Rename. Then type the file name in the Rename dialog box. You can also use the File menu to copy, move, and delete files.

These procedures for using a mouse to manipulate files also apply to directories. In addition, you can use the File menu to create a new directory and view more than one directory at a time. To create a directory, you first select the *parent directory,* which is the directory you want the new directory to reside in. Then from the File menu, choose Create Directory... Next, supply the directory name in the Create Directory dialog box. To change the current directory, simply use the mouse to select the desired drive and directory.

Copying and moving directories and files is facilitated by viewing more than one window at a time. From the Window menu, choose New Window. Then, you can organize the desktop by minimizing the windows. In addition, you can use the Window menu to tile the windows.

Try It Experiment with some of the File Manager features.

- Open the File Manager.
- Insert the MYDISK diskette into drive A.
- Display the tree and directory structure of the diskette in drive A.
- From the View menu, choose Tree and Directory.
- Open the TEMP directory.
- Open the COM directory.
- Use the View menu to display complete file information.
- Copy the EDIT.COM file from the COM directory to the HLP directory, and verify that there is a copy of EDIT.COM in the HLP directory by opening the HLP directory folder.
- Delete the EDIT.COM file in the HLP directory.

- Move the EDIT.COM file from the COM directory to the HLP directory, and verify that the move has taken place.
- Rename the EDIT.COM file to the name ED.COM.
- Create a new subdirectory called EXE in the COM directory, and verify the addition of the EXE directory folder under the COM parent directory.
- Remove the MYDISK diskette from drive A.

SUMMARY

This chapter introduced you to the Windows graphical user interface. You used a mouse to select objects in the Windows desktop. You then learned how to use the Program Manager to run and manage programs, and the File Manager to manipulate files and directories. You also learned how you can use online help to discover more about the numerous Windows features. In the next chapter, you will learn how to use several of the Windows accessories.

Key Words

applications window
buttons
Cascade
check box
click
command buttons
Control-menu box
desktop
dialog boxes
directory pane
directory structure
directory window
document windows
double-click
drag
drop-down list box
drop-down menu
File Manager
file pane
group window
insertion point
Maximize button
menu bar
Minimize button
mouse pointer
options button
parent directory
point
Program Manager
Restore button
scroll bar
scroll box
select
Task List
text box
Tile
title bar
windows

Exercises

1. Use Help in the Program Manager to learn how to organize applications.
2. Open the File Manager and copy MSAV.HLP and EDIT/HLP from the HLP subdirectory of the MYDISK diskette in drive A to the COM subdirectory of the MYDISK diskette in drive A.
3. Use the File Manager to sort by size all the .COM files in the COM subdirectory on the MYDISK diskette in drive A.

4. Create a directory called TEMP2 on the MYDISK diskette in drive A. Copy the EDIT.COM file from the COM subdirectory of the MYDISK diskette in drive A to the TEMP2 directory by using the File Manager. Rename this file EX.COM.
5. View the contents of the TEMP2 directory created in Exercise 4 on the diskette in drive A. Change the attribute of the EX.COM file in the TEMP2 directory so that the file is hidden.
6. Remove the TEMP2 directory created in Exercise 4 from the MYDISK diskette in drive A.

4 Windows Accessories

Software Development Cycle

Most computer users find Windows' graphical user interface an improvement over the command-line interface of DOS and are often surprised to learn about the range of useful accessories included in Windows. But Windows is not a fixed entity; new versions are designed to address user complaints and growing expectations. For example, the next version of Windows will replace the Program Manager and File Manager with an easier-to-use interface, provide shortcuts to access files or applications, avoid system crashes by shutting down problem applications, and enable you to use the right mouse button to access a context-sensitive menu that tells you what you can do next.

INTRODUCTION

Chapter 3 introduced the Windows desktop and demonstrated how to use Windows to manage files and programs. This chapter shows how to use six of Windows' powerful accessories: CardFile, Notepad, Calendar, Scientific Calculator, Paintbrush, and Write. You learn to create lists with Cardfile, to write short text files with the Notepad, to set appointments with the Calendar, and to perform calculations with the Scientific Calculator. In addition, you learn to create drawings on the screen with Paintbrush and to create and edit a document with Write.

4-1 CREATING LISTS WITH CARDFILE

Cardfile is a program that functions like a stack of index cards onto which you can write ideas, lists of things to do, and other information. These cards can be electronically managed (edited, copied, deleted, printed, and so on) and used to automatically dial selected telephone numbers.

Starting Cardfile

In the Accessories window, double-click the Cardfile icon. Maximize the Cardfile window so that it appears as shown in Figure 4-1.

The *status line* shows the current method for viewing the cards and the number of cards.

Adding a Card

You can add a card by choosing Card, selecting Add, and typing an index line in the Add dialog box. The *index line* is the text that appears at the top of the card and is used for identifying and sorting the cards. You can quickly add a card by pressing F7. To edit the index line, you select the card you want to change by clicking the index line. This card now appears in the front of the deck. To complete the index change, choose Edit, select Index, and type the text for the new index data.

Figure 4-1
Cardfile Window

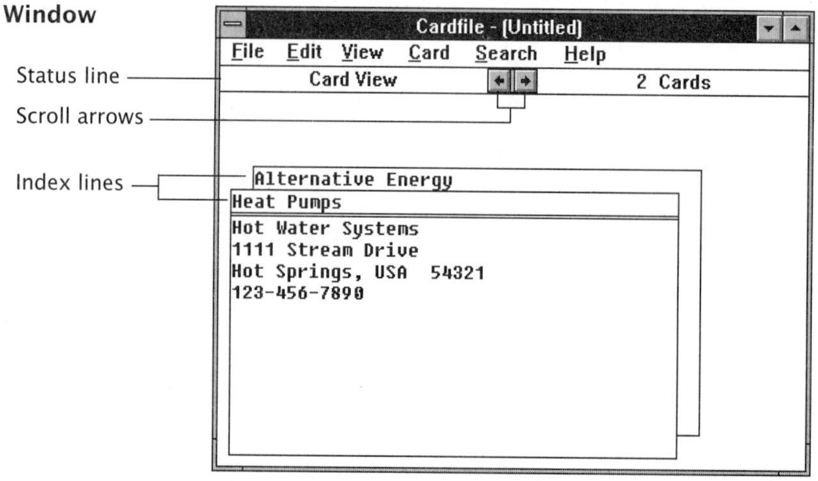

Saving a Cardfile

To save a Cardfile, choose File, select SaveAs..., and type the file name without typing an extension. Windows automaticaly appends `.crd` to the filename.

Viewing a Card

There are several ways to view the information on the cards. To scroll through each card in succession, choose View, select Card, and click the left or right scroll arrow, as shown in Figure 4-1. As you click a scroll arrow, the front card changes. If you hold down the left button on a scroll arrow, the index entries will revolve through the stack of cards.

To view one card in particular, choose View, select Card, and click the index line of the desired card.

You can view the cards as a list of index lines by choosing View and selecting List. To display a card on the top of the list, choose View, and select Card.

Searching for Text

You can search for key words or phrases in the text of a card (other than the index line) by choosing Search, selecting Find, and typing the word(s) to search for in the Find dialog box. The search proceeds up or down in the Cardfile depending on the direction selected. Uppercase is matched if you click the box next to the words *Match Case.* If a match is found, the card that contains the key word or phrase appears in the front. Found words are highlighted.

You can select Find Next to repeat the search for the same word(s).

Copying, Cutting, and Pasting

To copy text from a card to the Clipboard, you can drag the mouse to highlight the desired text. If you want to copy the text and still retain the original, choose Edit and select Copy. If you want to remove the text from the card, choose Edit and select Cut.

The text is now in the Clipboard and may be copied to an existing card or a new card. To copy the text you position the cursor at the new location, choose Edit, and select Paste. If you paste to an index line, only the first line of the Clipboard text is transferred.

Undoing

You can undo the last action by choosing Edit and selecting Undo. For example, if you select text in a card and then cut the text by choosing Edit and selecting Cut, the text disappears from the card. If your intent was to copy rather than cut, you can undo this last action of cutting by choosing Edit and selecting Undo.

While a card remains at the front of the list, you can make successive changes and still return to the version of the card you started with by choosing Edit and selecting Restore. Once the card has been moved from its front position, however, you cannot restore it.

Try It In this exercise, you use Cardfile to set up a resource list.

- Start Cardfile if it is not already running.
- Add a card with index line `Heat Pumps` containing the information

  ```
  Hot Water Systems
  1111 Steam Street
  Hot Springs, USA 54321
  123-456-7890
  ```

  by pressing F7, typing `Heat Pumps`, clicking OK, and typing the text.
- Edit the index line of this card to read `Geothermal Heat Pumps`.
- Save the Cardfile to a file named ENVIRO on the A drive.
- View the Cardfile as a list of index lines.
- Search for the word *water*.
- Display the `Cannot find` message by searching for the word *wood*.

4-2 USING THE NOTEPAD

Notepad is used for writing short- to medium-length text files. Since the text files created in Notepad have no special formatting characters such as carriage returns and linefeeds, Notepad is ideal for creating files to be read by other word processing programs.

Running Notepad

To run Notepad, double-click the Notepad icon in the Accessories group. You will see the Notepad window shown in Figure 4-2.

Figure 4-2
Notepad Window

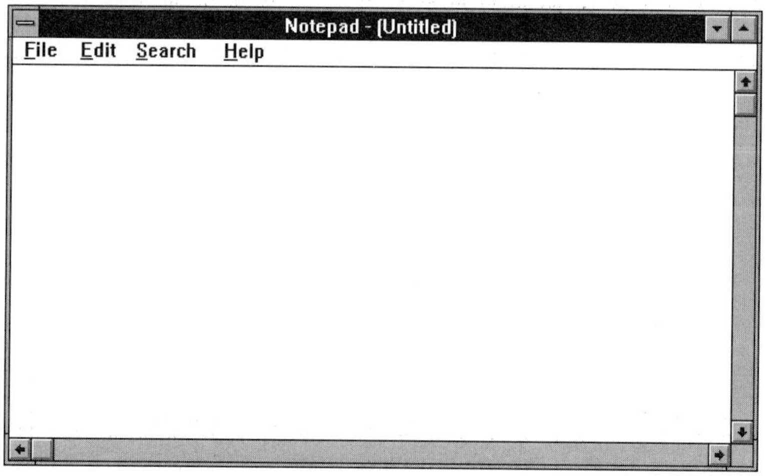

Entering and Editing Text

As you type, the text appears at the location of the insertion point on the screen. Continue typing, and the screen shifts so that you no longer see the beginning of the line. A more convenient way to view the text is with the words automatically wrapped to the beginning of the next line. To set this feature, choose Edit and select Word Wrap. Although the text is now more conveniently displayed, Notepad does not insert carriage return and line-feed characters into the text.

Four keyboard key combinations facilitate moving around in Notepad.

HOME	Moves the cursor to the beginning of a line
END	Moves the cursor to the end of a line
CTRL+HOME	Moves the cursor to the beginning of the document
CTRL+END	Moves the cursor to the end of the document

Selecting Text

In addition to selecting text by dragging the mouse to highlight the desired text, you can select the entire document by choosing Edit and selecting All. This makes it easier to copy the entire document into the Clipboard for transferring text to other documents.

Entering the Time and Date

To record when text is being written, choose Edit and select Time/Date. This information will appear in the Notepad document at the location of the cursor, as shown in Figure 4-3.

Figure 4-3
Time and Date

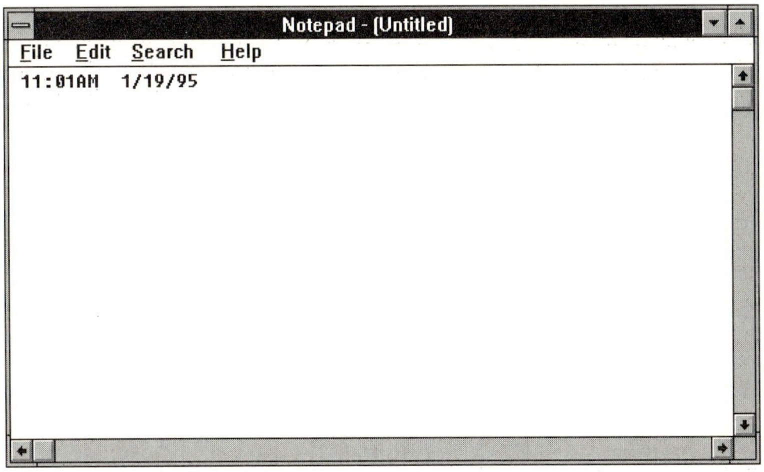

Searching for Text

To search for text, you choose Search and select Find, which opens the Find dialog box. Then, you enter the text to be found and the search direction (up or down) with respect to the current position of the insertion point. You can also select case-sensitive matches (where lowercase letters only match lowercase letters and capitals only match capitals). If you want to search again for the same key word(s), choose Find and select Next.

Managing Time

Notepad lets you keep track of how you spend your time. To create a time-log document, type `.LOG` at the top of a Notepad document. Then choose File and select Save. Each time you open this file to list your completed activity, the day and time are recorded.

Try It In this exercise you use various features of the Notepad application.

- Start the Notepad application if it is not already running.
- Without pressing ENTER, type the following text:

```
The Notepad program provides a convenient way to record
information for memos and reminders that can be merged
with other documents or maintained as they are.
```

- Choose Edit and select Word Wrap to view all the text.
- Position the cursor at the end of the document.
- Move the cursor to the next line.
- Insert the current day and time by choosing Edit and selecting Time/Date.
- Search for the word *maintained*. Use the up direction for the search.
- Close the Notepad application.

4-3 SETTING APPOINTMENTS WITH CALENDAR

Just like its paper counterpart, the *Calendar* application lets you schedule appointments and record important events. In addition, the Calendar has an alarm that can remind you of appointments.

Starting the Calendar

Double-click the Calendar icon in the Accessories group. You should see the window shown in Figure 4-4.

Entering and Removing Appointments

To enter a monthly appointment calendar, choose View and select Month. The month view shown in Figure 4-5 appears. The current month is displayed, and the current day has brackets on each side of the date. Click the right arrow button to move the month forward and click the left arrow button to move the month backward. To the right of the arrow buttons is the *appointment date.* This date is displayed in reverse video on the monthly calendar. To select a particular date without using the arrow buttons, choose Show, select Date, enter the date in the Show Date dialog box, and then click OK. The format choices for the date are month/day/year or month-day-year. If you want a year beyond 1999, you need to enter the full four characters for that year.

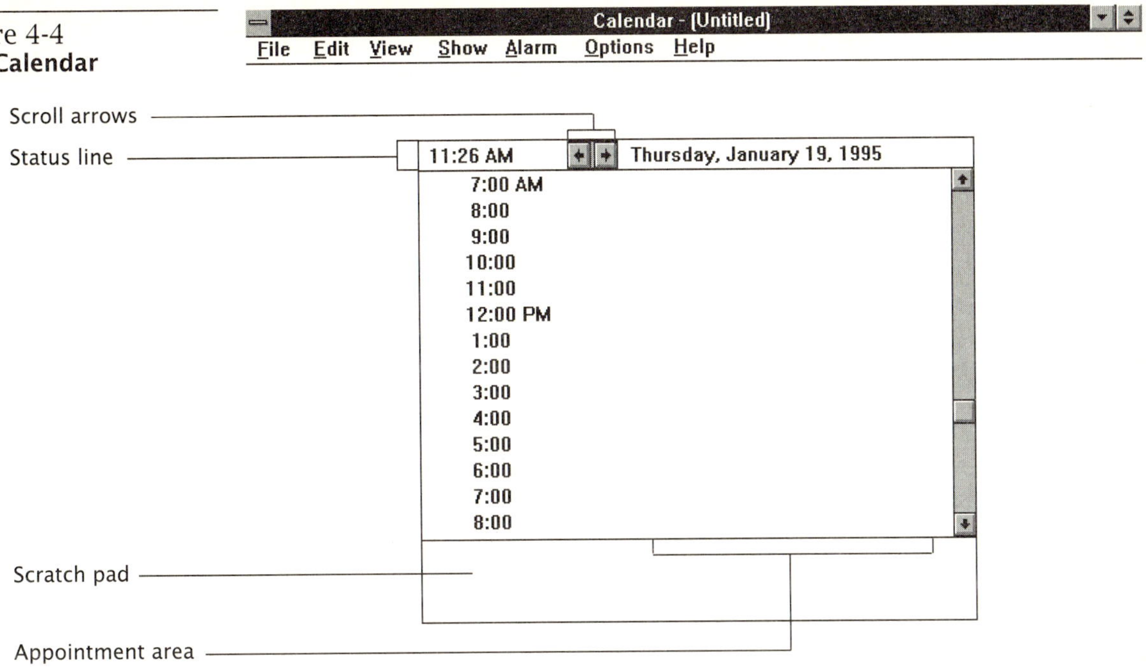

Figure 4-4
Day Calendar

To change the view to the day calendar either choose View and select Day or double-click the date that is to the right of the scrolling arrows in the monthly calendar. Double-clicking this date in one view of the calendar brings up the other view.

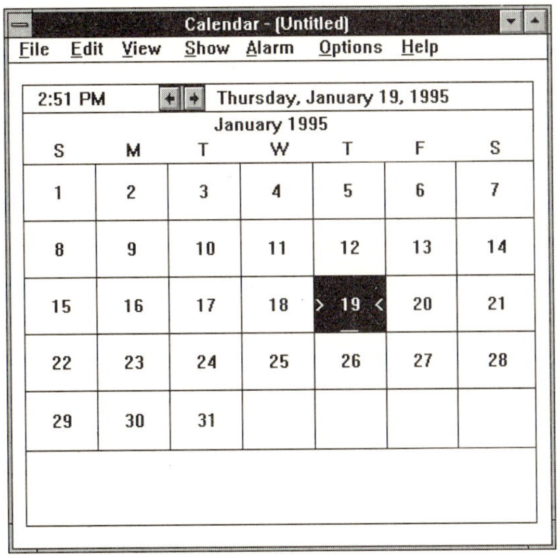

Figure 4-5
Month Calendar

To enter an appointment, position the cursor to the right of the time in the day calendar. Type in the relevant appointment information (who, where, what, and why) in the appointment area. You can modify the display times by choosing Options and selecting Day Settings. Then select 15-, 30-, or 60-minute intervals, a 12-hour or a 24-hour format, and a starting time other than the 7:00 a.m. default time.

You can add a special time to the day calendar by choosing Options and selecting Special Time. Then enter the time in hour:minute format, select AM or PM if using a 12-hour clock, and click Insert. To delete a special time, position the cursor to the right of the special time to be deleted. Then choose Options, select Special Time, and click Delete in the Special Time dialog box.

To mark a date with a symbol that reminds you of a relevant event like someone's birthday, you choose Options and select Mark to pick any one of five special *markings.* To remove a mark on a date, select the marked date in the monthly view. Then choose Options, select Mark, clear the check box on the mark for that date, and click OK.

To remove appointments from a day, choose Edit and select Remove. Then type in the From: date and the To: date, and click OK. If only one day's appointments are to be removed, leave the To: date empty.

At the bottom of the day view is a scratch pad for adding additional remarks.

Setting the Alarm

To set the alarm to sound at a particular time, display the day calendar and insert a special time. Then position the insertion point to the right of that time, choose Alarm, and select Set. A bell icon will appear to the left of the alarm time. You can remove a setting by positioning the insertion point to the right of a set time, choosing Alarm, and selecting Set. The bell disappears. When the actual time corresponds to this setting, four beeps will sound. If the Calendar application is active, the title bar will flash, and a dialog box will appear reminding you of the appointment. If the Calendar application is reduced to an icon, the icon will flash. Click once on the icon, and the reminder dialog box will appear. To turn off the sound or to cause the alarm to ring from 0 to 10 minutes early, choose Alarm and select Options.

Try It Use the Calendar application to schedule a typical day.

- Start the Calendar if it is not already running.
- Select tomorrow as the day to plan either by choosing View, selecting Month, and clicking the day square or by choosing View, selecting Day, and clicking the right scroll arrow.
- Enter an appointment to visit the tutor at 8:00 a.m. Use the text

  `Visit Tutor.`

- Select a special time of 10:15.
- Schedule the 10:15 appointment by positioning the insertion point next to the 10:15 time and typing

  `coffee break.`

- Set the alarm for the 10:15 appointment.
- Save this calendar by inserting a diskette in drive A, choosing File, selecting Save As, typing `A:TEMP.CAL`, and clicking OK.

4-4 PERFORMING CALCULATION WITH THE SCIENTIFIC CALCULATOR

In this section, you will use the Windows *Calculator* to perform calculations that include trigonometric functions and statistical operations. Though Windows provides two calculators, you will use only the Scientific Calculator in this section.

Running the Calculator

After double-clicking the Calculator icon in the Accessories group, open the Scientific Calculator by choosing View and selecting Scientific. You should see the window shown in Figure 4-6.

To enter numbers, you can use the keyboard, the numeric keypad (with NUM LOCK on), or the mouse to point to the Calculator buttons on the screen. For example, to multiply 15 times 16, click Dec (decimal) if it is not already on. Then click 1, click 5, click *, click 1, click 6, and click =. To clear an entry, click CE (clear entry). To start over, click C (clear) or press ESC. You can click BKSP to remove the last digit entered.

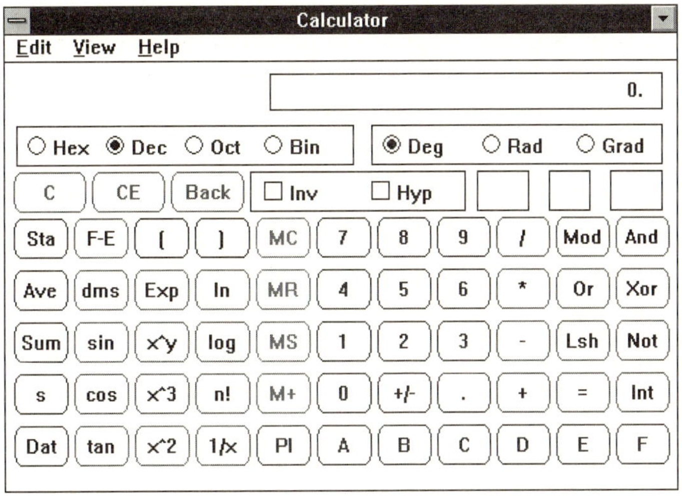

Figure 4-6
The Scientific Calculator

Using Memory

The Calculator application has its own *memory* to store numbers for later use. Choose MS (memory store) to store the displayed result in memory. Choose MR (memory recall) to display the result stored in memory. Choose MC (memory clear) to erase the result stored in memory. Choose M+ (memory add) to add the displayed number to the value stored in memory. For example, to add the first five positive integers, click Dec, click MC, click C, click 1, click MS. The M appears in the second box from the right indicating that a nonzero number resides in memory. Then click 2, click M+, click 3, click M+, click 4, click M+, click 5, and click M+. The sum of 1 through 5 is now stored in memory. To display this sum of 15, click MR.

Using the Clipboard

To save a displayed result on the *Clipboard*, choose Edit and select Copy. This result can then be pasted into other applications such as Cardfile and Notepad.

You can also paste from the Clipboard into the Calculator and produce a simplified macro, where *macro* is a command made up of a sequence of other commands. In this case, the Calculator responds to the characters in the Clipboard as either values or operations. For example, to add the first five positive integers, enter 1 + 2 + 3 + 4 + 5 = as text in Notepad. Then copy this text into the Clipboard, and paste this string into the Calculator (select Dec and click C before pasting). The sum of 15 appears in the display.

To use the Calculator memory in the Clipboard sequence, copy :c 1 :m 2 :p 3 :p 4 :p 5 :p onto the Clipboard from the Notepad application. The codes :c clears memory (MC), :m stores the displayed value (a 1) into memory (MS), and :p adds the displayed value into memory (M+). When you paste from the Clipboard into the Calculator, 5 appears in the display (the last number entered), and the M appears in the second box from the right indicating that there is a result in memory. Clicking MR displays the result of 15. To save time, you can copy to the Clipboard using CTRL+C and paste from the Clipboard using CTRL+V.

Using Different Number Systems

The Scientific Calculator supports Dec (decimal), Hex (hexadecimal or base 16), Oct (octal or base 8), and Bin (binary) numbers and operations. To convert a displayed number to one of the other three systems, simply click the desired button. For example, to convert decimal 10 to Hex, you first display 10 by clicking Dec, clicking 1, and clicking 0. To convert the displayed 10 to Hex, click the Hex button, and the display changes to A.

All four number systems accommodate positive and negative numbers, but only the Dec system accommodates nonintegers. For example, if you displayed –11.2 in Dec, the Hex conversion is F5, whereas F5 converted back to Dec is –11. The Scientific Calculator truncates decimal numbers to the integer format of Hex, Oct, and Bin.

You can easily perform operations normally reserved for decimal numbers on numbers in the other three systems. For example, to square 111 in binary, you click Bin, click 1, click 1, click 1, and click x^2. (The caret ^ symbol is used for exponentiation.) The displayed result of 11001 when converted to decimal is 49. Keep in mind that the other three number systems provide integer results only, and some operations are neither useful nor appropriate.

You can display numbers in their full 32-bit (binary digit) form by selecting Dword (double word). The lower 16 bits are displayed but not changed by selecting Word. To display the lower 8 bits of a number, again without changing the number, select Byte.

Using Statistics Functions

Here is a quick tour of the Scientific Calculator keyboard starting with the third row.

Sta (statistics) opens the Statistics Box for accessing the Ave (average), Sum (sum), s (standard deviation), and Dat (data) functions. The Statistics Box is shown in Figure 4-7.

Figure 4-7
Statistics Box

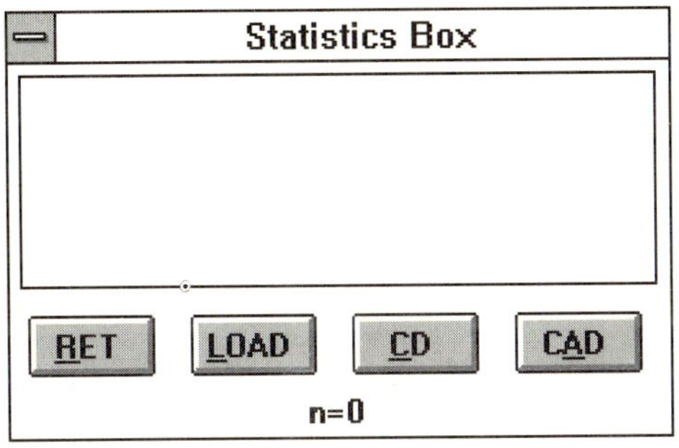

For example, you can compute the sum, mean, and standard deviation of the first five integers by clicking Sta, which opens the Statistics Box. Then, drag the Statistics Box over to the side so that you see both this box and the Calculator. If you are using the keyboard and not the mouse, you click the Ret (return) button in the Statistics Box to return to the calculator. Then, you click 1 to display the first number, and click Dat to enter this value into the Statistics Box. Click 2 to enter this value into the Statistics Box. Continue this procedure for the numbers 3, 4, and 5. If the display of numbers is not responding, make sure that the Dec number system is selected.

You can delete the highlighted data in the Statistics Box by clicking CD (clear data). The highlight can be moved by clicking the desired data. Selecting CAD (clear all data) removes all the current data from the Statistics Box. To change the number in the Calculator display to the number highlighted in the Statistics Box, click Load. Thus, Load and Dat are used for transferring values between the Calculator and the Statistics Box.

You can compute the mean of the numbers displayed in the Statistics Box by clicking Ave. The result of 3 appears in the Calculator display. The sum of 15 is displayed by clicking Sum, and the standard deviation is displayed by clicking s. To change the population parameter in the standard deviation from n–1 to n, select Inv prior to selecting s.

To calculate the sum of the squares, select Inv and then select Sum. Selecting Inv and then selecting Ave calculates the mean of the squares.

F-E switches between normal display and exponential (scientific) display for decimal numbers. For example, you could enter 256 and click F-E to toggle between the scientific display of 2.56e+002 and the display of 256.

You can use dms to display the decimal number of degrees in terms of degrees, minutes, and seconds. For example, if you click C, click Dec. and enter 15.21, this number represents 15 degrees plus .21 times 60 = 12.6 minutes. You could click dms and observe 15.1236. That is, 15.21 degrees is 15 degrees, 12 minutes, and 36 seconds. To convert back to degrees, click Inv and then click dms.

Using Trigonometric Functions

You can use sin, cos, and tan to calculate the respective trigonometric functions of the displayed number depending on whether the number is in Deg (degrees), Rad (radians), or Grad (gradiens). You can select Inv or Hyp (hyperbolic) to calculate such trigonometric quantities as the inverse hyperbolic tangent and the inverse cosine.

You can use the left and right parentheses to open and close up to 25 levels of computation where the innermost level is calculated first.

Using Mathematical Functions

To enter the exponent portion of a decimal number, you use Exp. The range of this exponent is from -307 to +307.

You can use x^y to calculate x to the y power. Preselecting Inv calculates the yth root of x. For example, you can find the square root of 49 by displaying 49, selecting Inv, selecting x^y, selecting 2, and selecting =, which displays 7.

To square a displayed number, use x^2. Preselecting Inv calculates the square root. Similarly, x^3 cubes the displayed number. Preselecting Inv calculates the cube root.

You can calculate the natural logarithm of the displayed value by selecting ln. Preselecting Inv calculates e to the power of the displayed value. To calculate the base 10 logarithm of the displayed value, use log. Preselecting Inv calculates 10 to the power of the displayed value.

Int displays the integer portion of the displayed number. For example, if the displayed number is 12.3, Int displays 12. Selecting Inv and then selecting Int displays the fractional portion of the number. In this example, Inv Int displays .3.

Additional mathematical functions are available. n! calculates the factorial of the displayed value. 1/x calculates the reciprocal of the displayed value. PI displays the value of π, whereas preselecting Inv displays 2p. You also can calculate the modulus of the displayed number with respect to a given base by using Mod. For example, keying 10 Mod 4 displays 2.

You can use on-line Help to learn about additional features of the Scientific Calculator.

Try It In this exercise, you use the Scientific Calculator.

- Start the Scientific Calculator if it is not already running.
- Use the Calculator memory to store the number 46 and to add 3.
- Copy the displayed number 49 onto the Clipboard.
- Run the Cardfile application and paste the number 49 from the Clipboard.
- Enter the following Calculator sequence on a card in the Cardfile, copy it to the Clipboard, and run it on the Calculator by pasting it: (1 + 5 + 6)/3 =.
- Convert the decimal number 5564 to binary, octal, and hex.
- Compute the square root of the sum of the squares of the numbers -4, 3, 1.
- Store the result of the previous calculation in memory, and display the integer part.

4-5 USING PAINTBRUSH

Paintbrush lets you create drawings on the screen in a manner similar to painting on a canvas. You select colors and tools to draw images that can be printed or used in other applications.

Starting Paintbrush

Double-click the Paintbrush icon in the Accessories group. You should see a screen similar to the one shown in Figure 4-8.

Figure 4-8
Paintbrush Work Area

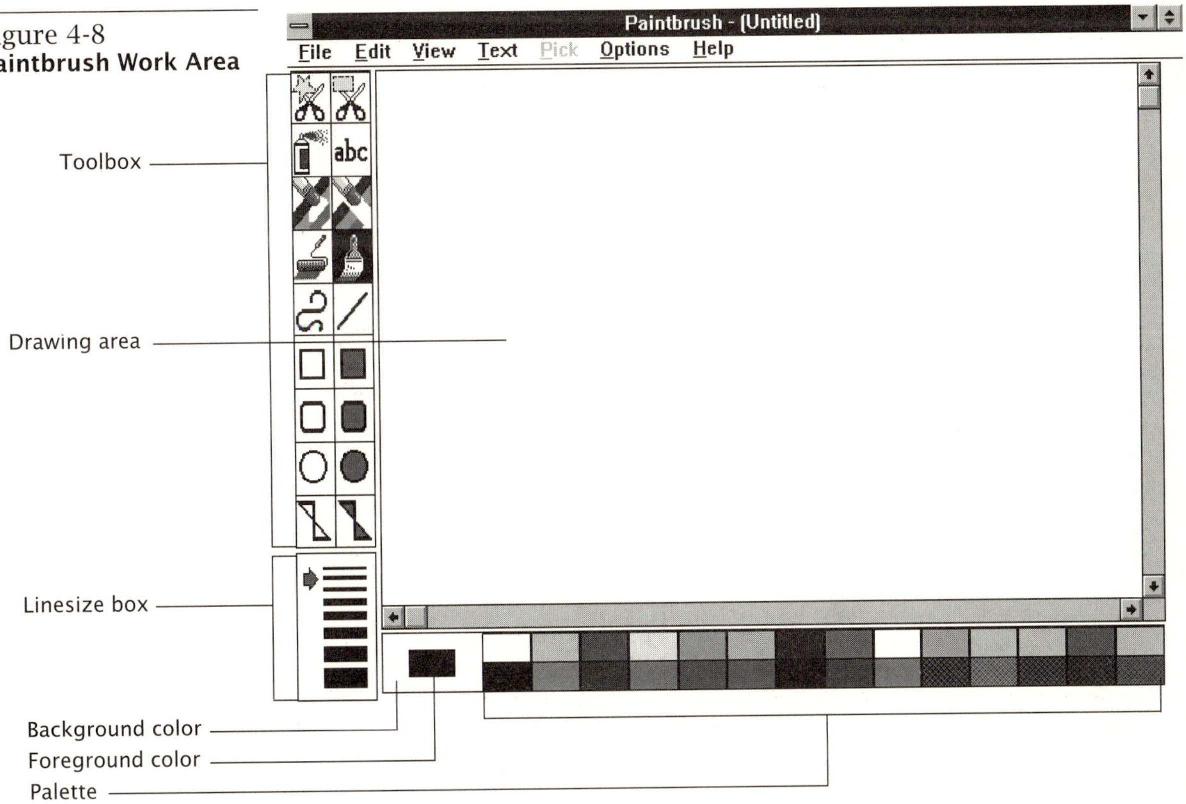

On the left-hand side is a *Toolbox* that contains drawing tools that help you draw, fill, airbrush, erase, enter text, and select items in the drawing area. On the right side is the *drawing area*, which is where you create your illustrations.

Changing the Drawing Size

Paintbrush determines a default drawing size based on the amount of available memory and the type of display adapter in use. To see this size and to change it, choose Options and select Image Attributes.... This displays the Image Attributes dialog box. The units of in (inches), cm (centimeters), and pels (picture elements or pixels) are associated with the numbers you select for the width (width of the drawing) and height (height of the drawing). You can select black and white or colors for the drawing. If you select OK, the new changes are in effect. If you choose

Default, the drawing size that is based on the available memory and display adapter is in effect. If you choose Cancel, no change is made to the drawing attributes.

Below the Toolbox is a *Linesize box* for selecting drawing widths. At the bottom of the window to the right of the Linesize box is a *Palette* of colors and patterns used for the background and the drawing tools.

Increasing the Viewing Area

To see more of the drawing, you can remove the Toolbox, the Linesize box, and Palette. Choose View, and select Tools and Linesize. Then choose View again and select Palette. You can display these items by choosing View and selecting Tools and Linesize. Then choose View again and select Palette.

You can see still more of the drawing by choosing View and selecting View Picture. This shows as much of the drawing as will fit on the screen. To continue working on the drawing, click anywhere on the screen.

Choosing View and selecting Zoom In causes Paintbrush to shrink the entire drawing to fit on the screen. In this view, you can cut, copy, and paste, but you cannot add anything new to the drawing. To continue developing the drawing, choose View and select Zoom Out.

Selecting Color and a Drawing Width

The background color is the color of the canvas, and the *foreground color* is the color of the paint used on the brushes. On the left of the Palette is a rectangle within a rectangle. The outer rectangle is the *background color*, and the inner rectangle is the foreground color. To change the background color, point the mouse cursor to the desired color in the Palette and click the right mouse button. You select the foreground color with the left mouse button. As you do this, note the corresponding changes to the rectangle within a rectangle.

You can select a width for the tools by clicking the desired line width in the Linesize box. This width has an effect on most of the tools in the Toolbox.

Using the Toolbox

To select a tool, click its icon in the Toolbox shown in Figure 4-9. The tools work in conjunction with the line width selected in the Linesize box and the foreground and background colors selected in the Palette.

Cutting Portions of the Drawing

You can cut out the desired portion of your drawing by using scissors or pick. Scissors creates a freehand type of enclosure that is drawn as you drag the cursor. Pick creates a more precise rectangular box. If you are unhappy with the resulting enclosure, delete it by clicking anywhere in the drawing. Once a portion of the drawing has been cut out, you can use the Edit menu to cut, copy, and paste the portion.

Figure 4-9
The Toolbox

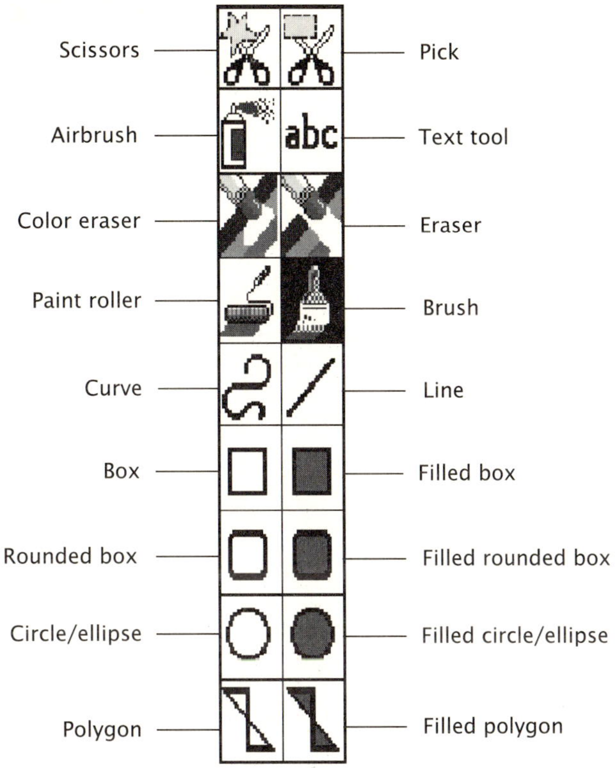

Airbrushing

You can also select airbrush and move the cursor into the drawing area. As you drag the mouse, a circular spray of paint covers the canvas with a density that depends on how fast you drag the mouse.

Writing Text

You can add text by choosing Text, selecting Fonts..., and completing the Font dialog box. As you choose different Fonts, Font Styles, Sizes, and Effects, you see a sample of the resulting text displayed in the dialog box. Click OK or Cancel to return to the drawing.

You can select Text and click at a location in the drawing where you want the text to start. As you type from the keyboard, the text appears in the drawing but does not wrap around. You can use BKSP to alter the text in a limited way until you have selected another tool. Once the text is in place, you can remove it by cutting it out of the drawing or erasing it.

Erasing

Both color eraser and eraser depend on the line width selection in the Linesize box. Color eraser erases the current foreground color and replaces it with the current background color. Eraser (to the right of the color eraser) replaces all colors with the current background color. To move either tool in orthogonal motions, you hold down SHFT as you drag the mouse.

Paint Rolling

To fill in enclosed areas, you use paint roller. An *enclosed area* is defined by at least three connecting lines or curves. When you click the area to be filled, some of the other areas might be filled in as well. If this effect is undesirable, choose Edit and select Undo.

Brushing

Use brush to draw freehand on the canvas. You can use different brush shapes by choosing Options and selecting Brush Shapes. Then select one of the shapes in the Brush Shapes dialog box and click OK. To brush vertical and horizontal lines, hold down SHFT while dragging the mouse.

Drawing Curves

To draw curves, you first select the line width and color, and then select the curve. Next use the mouse to position the cursor, and click to anchor one end of the curve. Then, drag the cursor to where you want the curve to end. Next drag the cursor and the line will arch in the direction of the movement. If this curve satisfies your requirements, click the second endpoint. Otherwise, drag the cursor again to bend the line once more. Once you release the mouse button this second time, the curve is drawn.

Drawing Lines

You can draw a line by selecting line and clicking and dragging the mouse within the drawing area. Once you release the mouse button, the line is drawn. If you want to erase the line, you can click the right mouse button at any time that the left mouse button is still pressed. To draw horizontal, vertical, and 45-degree lines, hold down SHFT while dragging the mouse. You can see the cursor coordinates, choose View and select Cursor Position.

Using the Box and the Filled Box

Selecting box lets you draw boxes whose outlines are the current foreground color. By selecting filled box, you can draw boxes whose outlines are the current background color and whose filled interiors are the current foreground color. To display no border, make the background and the foreground colors the same. To draw a perfect square (rather than a box), hold down SHFT while dragging the mouse. As with the line tool, you can delete a box by clicking the right mouse button at any time while the left mouse button is still pressed.

Using the Rounded Box and the Filled Rounded Box

To create boxes with rounded corners, use rounded box and filled rounded box. These icons function in the same manner as box and filled box except that the corners are rounded.

Using the Circle/Ellipse and the Filled Circle/Ellipse

To draw an ellipse with an outline that matches the current foreground color, choose circle, and select ellipse. Then position the cursor in the drawing area. As before, you can delete the ellipse by clicking the right mouse button while the left mouse button is still pressed. You also can draw circles by holding down SHFT while dragging the mouse.

Filled circle/ellipse is similar to filled box in that the outline takes the current background color, and the fill color is the current foreground color.

Using the Polygon and the Filled Polygon

To draw a polygon, select polygon and click in the drawing area. You can click or drag to draw many-sided figures. Double-click to close the drawing. Hold down SHFT while dragging the mouse to draw vertical, horizontal, and 45-degree lines. Filled polygon functions in the same manner as polygon except that the figure is filled.

Try It

In this exercise, you use several Paintbrush tools.

- Start the Paintbrush if it is not already running.
- Create a drawing similar to the one shown in Figure 4-10 using the following tools: box, circle/ellipse, paint roller, brush, and text.

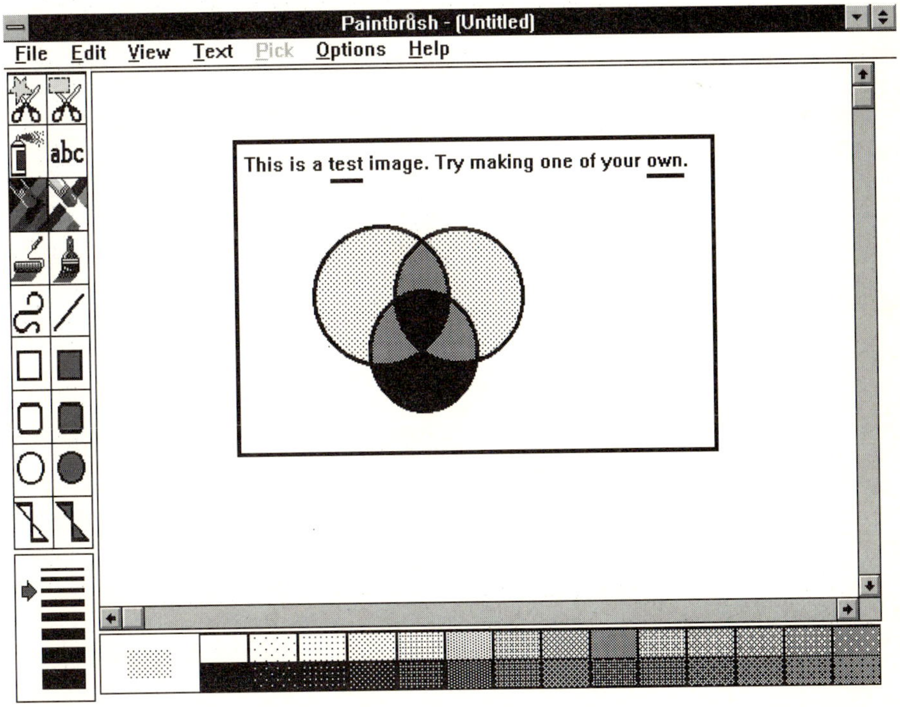

Figure 4-10
A Test Image

4-6 EDITING A DOCUMENT WITH WRITE

Windows supplies a modest word processing application called *Write* that can satisfy many of your needs for producing documents. In this section, you will be introduced to the features of this extremely useful program.

Starting Write

To start Write, double-click the Write icon in the Accessories group. You will see the screen shown in Figure 4-11.

Figure 4-11
The Write Window

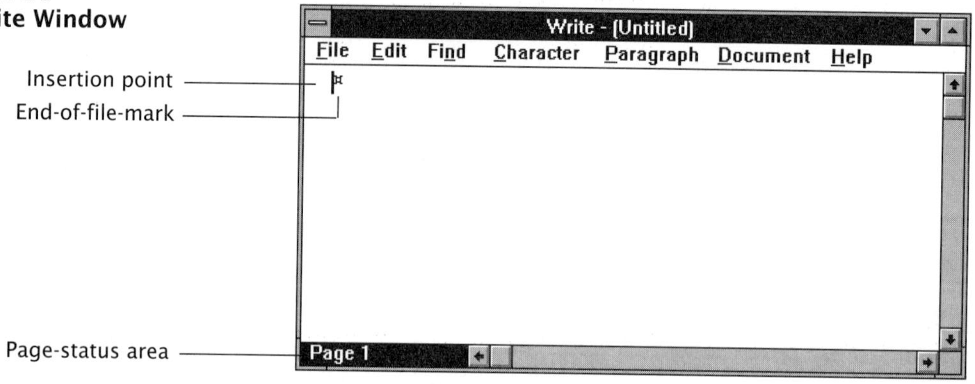

If you type `This is a test of Write` and press (HOME), you see a blinking vertical line in the top-left corner of the window. This line, called the insertion point, marks the place where the typed text will be inserted. You can move the insertion point with the keyboard arrow keys or with the mouse. Next, position the insertion point between the words *a* and *test*, type `quick`, and press (SPACE).

An *end-of-file* mark is a small symbol that marks the end of your document. At the bottom of the window is the page-status area, which tells you which page of the document you are currently on.

To enter more text in the document, reposition the insertion point and type. Write automatically wraps the text to the next line. To start a new paragraph, press (ENTER). To leave a line space, press (ENTER) twice.

Moving Around a Document

As your document becomes larger, it is useful to be able to quickly move to various places in the document. In addition to clicking the scroll bars, you can use the keyboard. To move to the beginning of a line, you can press (HOME), and to move to the end of a line, you can press (END). You can move to the end of the document by pressing (CTRL)+(END) and move to the beginning of the document by pressing (CTRL)+(HOME). To move to the next screen, press (PGDN) while to move to the previous screen, press (PGUP).

As you might expect, Write has extended movement features. These include (CTRL)+(PGDN), which moves to the bottom of the window, and (CTRL)+(PGUP), which moves to the top of the window. (CTRL)+(→) moves the insertion point to the next word, and (CTRL)+(←) moves the insertion point to the previous word.

You can divide the document into pages by choosing File, selecting Repaginate, and clicking OK in the Repaginate Document dialog box. You can move to the next page by pressing (5)+(PGDN) and move to the previous page by pressing (5)+(PGUP). If you hold down any of the previous key combinations, the action repeats.

Selecting Text

Selecting text means to highlight it so that it can be moved, copied, or changed as a block. The simplest way to select text is to press the left mouse button at the beginning of the text to be selected. This defines the *anchor*. Then, you can drag the mouse to the end of the block and release the mouse button. To deselect a block of text, click anywhere in the document window.

Here are some other methods for selecting text. To select a single word, double-click the word. To select an entire sentence, press and hold (CTRL), and then click any word in the desired sentence. You can increase the selected block one sentence at a time by dragging the mouse as you hold down (CTRL).

The selection area is to the left of the document. Moving the mouse pointer there causes the pointer to change into an arrow. To select an entire line of text, click next to the line in the selection area. To select more than one line at a time, drag the mouse in the selection area.

To select an entire paragraph, double-click in the selection area next to the desired paragraph. If you double-click and then drag the mouse down the selection area, you can select whole paragraphs at a time.

To select the entire document, place the mouse cursor in the selection area, hold down (CTRL), and click.

Editing Text

You can remove characters from the document by positioning the insertion point in front of the character and pressing (DEL) or by positioning the insertion point in back of the character and pressing (BKSP).

To remove a block of selected text, press (DEL). If you decide that this last deletion is a mistake, you can undo it by choosing Edit and selecting Undo. There is a quick way to replace a word in Write. You select the word, and then simply type in the replacement word. The highlighted word is deleted automatically, and the new characters are inserted.

Finding and Replacing Text

Sometimes you want to locate a portion of text in a document and then replace it with new text. To find text, choose Find and then select Find.... Next type the text to be found in the Find dialog box and press (ENTER). You have the option of matching the whole word or the case.

To automatically replace text that is found, choose Find and select Replace.... Then type the text in the Replace dialog box, and select either Find Next (to find the next occurrence of the desired text), Replace (to replace that next occurrence of the text with the replacement text), Replace All (to replace all occurrences of the search text with the replacement text), or Cancel (to end the replace operation).

Formatting the Document

Write provides a default formatting for your document. These defaults will likely be adequate for most of your work. If you want to enhance the appearance of your documents, however, you will need to use some of Write's additional features.

Using the Ruler

If you choose Document and select Ruler On, the ruler and icons shown in Figure 4-12 will appear. The left-indent marker appears as a dot within a triangle on the left side of the ruler. The dot represents the left indentation of the first line of each paragraph. The triangle represents the left indentation of all other lines. The right-indent marker controls the right margin of the document. You can select the entire document by pressing CTRL and clicking in the selection area. By dragging the left-indent marker, you can observe the changes in the left margin for all paragraphs. Dragging the dot changes the left indentation of the first line. By dragging the right-indent marker, you can change the right margin for all paragraphs.

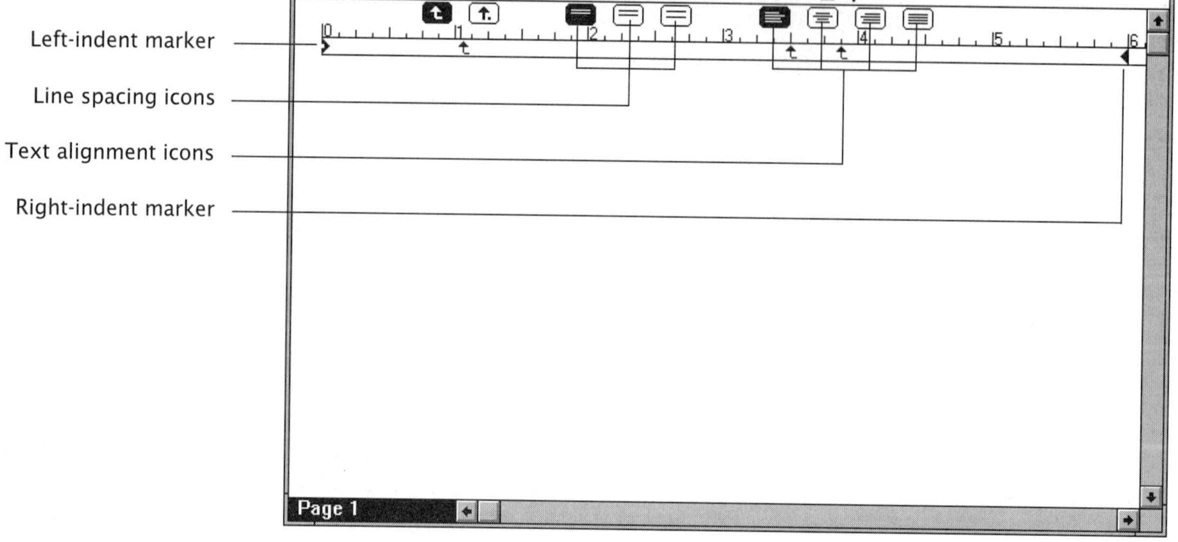

Figure 4-12
The Write Ruler

There are three icons on the ruler for line spacing: single spacing, one-and-a-half line spacing, and double spacing. The choice for alignment includes left justification, centered, right justification, and both left and right justification. To return to the normal format of single spacing, left justification, and zero indents, choose Paragraph and select Normal.

There are two types of tabs: left-aligned tabs and decimal-aligned tabs. Left-aligned tabs line up text on the left of the tab. Decimal-aligned tabs line up text at the decimal point that appears in a number. This feature is useful for making columns of monetary values. You can set up to a maximum of 12 tabs either by choosing Document, selecting Tab, and completing the Tabs dialog box. You can selectively clear tabs by double-clicking in the Position box of the tab you want to clear and then pressing DEL. To remove all the tabs, select Clear All. Figure 4-13 shows the use of tabs to create a cost table.

You can also use the mouse to click one of the tab icons and click the location on the ruler that you want the tab to appear. You can drag the tab marking to new locations as you wish. To clear a tab, click the tab and drag it into the document. The tab marking will disappear.

Figure 4-13
Use of Left-Aligned and Decimal-Aligned Tabs

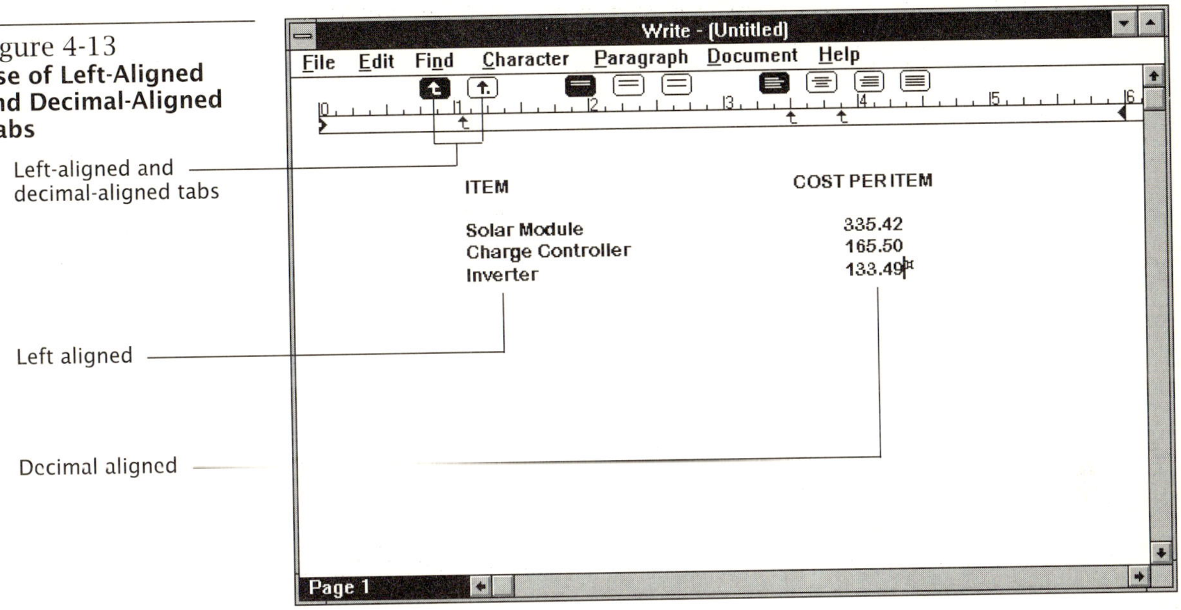

Changing the Page Layout

To personalize your document, you can add information in the header or footer of each page. A *header* is a line of text that appears at the top of a page and might display the title. A *footer* appears at the bottom of the page and might include the page number. To create headers and footers, you can choose Document, and select Header or Footer. Enter the desired text. You have the options of omitting the text on the first page, inserting page numbers, and specifying the distance from the top (or bottom). Then select Return to Document. Note that headers and footers appear only when you print the document.

To vary the appearance of the characters in the document, you can change the character *fonts* which are graphic designs for the characters you type at the keyboard. You can change the size of the selected text by choosing Character, and selecting either Reduce Font or Enlarge Font.

You can also change the style of the characters by making them bold, italic, or underlined. First choose Character. Then select Bold for boldface, Italic for italics, or Underline for underlining the selected text. You also can add subscripts or superscripts, by first choosing Character. Then select Subscript or Superscript.

For more control of the available fonts, choose Character and select Fonts.... Then select the Font, Font Style, and Size from the Font dialog box. If possible, Write prints the same fonts on paper that you see on the screen.

You can change preset margins and page numbering. Choose Document and select Page Layout. Then fill in the Page Layout dialog box.

Try It

In this exercise, you will create a document using Write.

- Start Write if it is not already running.
- Use Write to create a résumé. Include your academic as well as your employment record.
- Save the document by choosing File and selecting Save As. Type A: RESUME and check OK.

SUMMARY

This chapter introduced you to the Windows accessories Cardfile, Notepad, Calendar, Calculator, Paintbrush, and Write.

Key Words

anchor
appointment date
background color
Calculator
Calendar
Cardfile
Clipboard
Drawing area
enclosed area
end-of-file mark
fonts
footer
foreground color
header
index line
Linesize box
macro
markings
Match Case
memory
Notepad
Paintbrush
Palette
population parameter
selecting text
status line
Toolbox
Write

Exercises

1. Use Cardfile to store the following textbook information:

   Index: Robotics, Title: Mobile Robots, Author(s): Jones and Flynn, Date: 1993.
2. Copy the Title: Mobile Robots line from the card created in Exercise 1. Add a new card with Index: Books on Loan, and paste the Title: Mobile Robots line onto that card.
3. Use Notepad to create a memo to all engineering majors from the students of this course offering to help other students learn about Windows accessories.
4. Set the alarm in the Calendar application to go off in 10 minutes.

5. Use the Scientific Calculator to calculate the following:
   (a) log(ln e)
   (b) sin(cos 20°)
   (c) sinh(20°)
   (d) $\tanh^{-1}(.95)$
6. Use the Scientific Calculator to calculate the following:
   (a) $\sqrt{(-1)}$
   (b) 3π
   (c) $771^8/26^8$
   (d) $\sqrt{(771^8)}$
7. Use the Calendar to record your daily class schedule for one week. Save this information on a diskette in drive A.
8. Identify books in the bookstore that you feel would be useful for next year's freshman students. Be as thorough as possible.
9. Use Write to create a document descring how you selected the books you identified in Exercise 8.
10. Use Cardfile to produce a table of book information for Exercise 9. Include author, publisher, title, publication date, and cost. Use the Clipboard to transfer this information from the Cardfile to the Write document you created in Exercise 9.
11. Use Paintbrush to produce a pie chart that represents the relative costs of the books you identified in Exercise 10. Include this chart as a figure in the document you prepared in Exercise 9.
12. Use the Scientific Calculator to compute the total and the average cost for the books you identified in Exercise 10. Use the clipboard to transfer this information from the Cardfile to the Write document you created in Exercise 9.

Index

NOTES

NOTES

NOTES

NOTES

NOTES

NOTES

NOTES

NOTES

NOTES

NOTES

NOTES

NOTES